The Minerals, Metals & Materials Series

Antoine Allanore · Guillaume Lambotte
Jonghyun Lee
Editors

Materials Processing Fundamentals 2017

TMS

Springer

Editors
Antoine Allanore
Massachusetts Institute of Technology
Cambridge, MA
USA

Guillaume Lambotte
Boston Electrometallurgical Corporation
Woburn, MA
USA

Jonghyun Lee
Department of Mechanical and Industrial Engineering
University of Massachusetts
Amherst, MA
USA

ISSN 2367-1181 ISSN 2367-1696 (electronic)
The Minerals, Metals & Materials Series
ISBN 978-3-319-51579-3 ISBN 978-3-319-51580-9 (eBook)
DOI 10.1007/978-3-319-51580-9
TMS owns copyright; Springer has full administrative rights

Library of Congress Control Number: 2016961670

Printed on acid-free paper

This Springer imprint is published by Springer Nature
The registered company is Springer International Publishing AG
The registered company address is: Gewerbestrasse 11, 6330 Cham, Switzerland

Preface

The symposium Materials Processing Fundamentals is annually hosted at The Minerals, Metals & Materials Society (TMS) annual meeting, as the flagship symposium of the Process Technology and Modeling Committee. It is a unique opportunity for interdisciplinary presentations and discussions about processing across materials, across state of matter, and as always at TMS, across continents. The presentations and papers have a common theme: the quest for modeling—either numerical or physical—or sensing as applied to production or synthesis processes. The materials covered include ferrous and non-ferrous elements, and the processes range from mining unit operations to joining and surface finishing of materials. Indeed, recognizing that modern processes involve multi physics, the symposium and its proceedings allow novel or existing practitioners to learn the methods and outcome of other fields' modeling practices, often enabling the development of practical solutions to common problems. Clearly, modeling of basic thermodynamic and physical properties play a key role, along with computer fluid dynamics and multiphase transport and interface modeling. Contributions in this opus include applications such as steel secondary metallurgy, with developments for specific unit operations (continuous casting, ladle refining, slabs cooling, etc). Gaseous or molten state processes are studied, for example chemical properties of ferrous and non-ferrous systems (silicates, calcium chloride, sodium hydroxide, sulfides), of molten zirconium or gaseous gallium. Two models of mechanical and electrochemical processes are included, completing a broad view of the field and practices of modeling in materials processing. The engagement of the TMS members with the committee to chair sessions, review manuscripts, and help TMS to reflect the existing practices, makes such a symposium and its proceedings possible. The editor and its co-editors therefore thank warmly those volunteers along with the TMS staff, encompassing Patricia Warren and Matt Baker, for there exceptional support to allow the present opus to exist.

Antoine Allanore
Guillaume Lambotte
Jonghyun Lee

Contents

About the Editors

Antoine Allanore received his higher education in Nancy (France) where he earned a chemical process engineer diploma from Ecole Nationale Supérieure des Industries Chimiques and an M.Sc. and a Ph.D. from Lorraine University. Dr. Allanore joined the Massachusetts Institute of Technology (USA) in 2010 as a postdoctoral fellow, after several years of service as a research engineer for ArcelorMittal. In 2012, he was appointed the T.B. King Assistant Professor of Metallurgy in the Department of Materials Science & Engineering at MIT, where his research group aims at developing sustainable materials extraction and manufacturing processes. In particular, his laboratory investigates the electrochemical properties of metals and molten compounds, e.g. oxides and sulfides. He teaches thermodynamics and sustainable chemical metallurgy at both undergraduate and graduate levels. He received the Vittorio de Nora Award from The Minerals, Metals & Materials Society (TMS) in 2012 along with Dr. J. Yurko, and the TMS Early Career Faculty Fellow Award in 2015. Professor Allanore is the current Chair of the TMS Process Technology and Modeling Committee.

Guillaume Lambotte is a senior research & development engineer at Boston Electrometallurgical Corporation, a Massachusetts Institute of Technology (MIT) spin-off startup focusing on the development of an environmentally friendly and energetically efficient primary metal extraction process. Dr. Lambotte primarily focuses on computational thermodynamic modeling, electrochemistry and high-temperature equilibrium. Prior to joining BEMC, he conducted research as a postdoctoral associate at the University of Massachusetts (UMASS) Amherst and MIT. Before his graduate studies, Dr. Lambotte worked as a production assistant manager at Alcan Extruded Products (Crailsheim, Germany).

Dr. Lambotte obtained his bachelor degree from the European Engineer School for Materials Science (Nancy, France). He received an M.Sc. and a Ph.D. in Metallurgical Engineering from Ecole Polytechnique of Montreal (Montreal, Canada).

Dr. Lambotte is currently serving on the TMS Process Technology and Modeling Committee and was the recipient of the 2015 TMS EPD Young Leaders Professional Development Award. In 2015, Dr. Lambotte was one of the TMS representatives at the Emerging Leaders Alliance Conference.

Jonghyun Lee is a research assistant professor in the Department of Mechanical and Industrial Engineering at the University of Massachusetts Amherst. He has been conducting multiple industry- and government-funded projects in the field of materials processing as PI and Co-I. Dr. Lee is a recipient of the Young Leaders Professional Development Award in 2013 from The Minerals, Metals & Materials Society, where he has been serving as a co-organizer and co-editor of the Materials Processing Fundamentals Symposium since 2014. Prior to joining the current institution, he had 4.5 years of industry experience and worked as a postdoctoral associate for Tufts University, Medford, Massachusetts. He earned his M.S. and Ph.D. in Mechanical Engineering from the University of Massachusetts Amherst and his B.S. in the same discipline from Inha University, South Korea.

Part I
Steelmaking

Development of Ultrahigh-Basicity Mold Fluxes for Peritectic Steel Continuous Casting

Xiao Long, Shengping He, Qian Wang and P. Chris Pistorius

Abstract In the continuous casting of peritectic steels, mold fluxes play significant roles in providing defect-free surfaces of slabs. In this work, a traditional high-basicity mold flux was studied. Slag film samples were solidified from molten flux, with an improved water-cooled copper probe and using different probe immersion times. The film microstructures, interfacial roughness and porosity were studied. Based on the results, ultrahigh-basicity mold fluxes (CaO wt%/SiO_2 wt% = 1.5 ~ 1.8) were developed. A typical ultrahigh-basicity mold flux (CaO wt %/SiO_2 wt% = 1.74) was analyzed in the laboratory and applied in industrial processes. The laboratory experiments and industrial results indicate that the novel mold fluxes developed in this work have excellent performance in balancing heat transfer control by the solid film and lubrication by molten flux.

Keywords Mold flux · Peritectic steel · Solid slag film · Ultrahigh basicity

Introduction

Mold fluxes are widely used in the continuous casting of steels to improve the surface quality of slabs and the efficiency of casting [1]. In the production of peritectic steel slabs, the volume shrinkage caused by the peritectic reaction in the cooling of the initial steel shells can lead to longitudinal cracks on the surface and sub-surface of slabs. The usual method to decrease the cracking tendency on the slab surface is to form a less intensive cooling environment for initial shells near the meniscus [2]. Thus, the mold flux adopted in peritectic steel casting must provide

X. Long (✉) · S. He · Q. Wang
College of Materials Science and Engineering,
Chongqing University, Chongqing 400044, China
e-mail: xiaolong@cqu.edu.cn

X. Long · P.C. Pistorius
Department of Materials Science and Engineering, Carnegie Mellon University,
Pittsburgh, PA 15213, USA

A. Allanore et al. (eds.), *Materials Processing Fundamentals 2017*,
The Minerals, Metals & Materials Series, DOI 10.1007/978-3-319-51580-9_1

sufficient retardation of heat transfer through the slag film (the thermal resistance between the steel shell and the water-cooled copper mold). In the industrial process of peritectic steel continuous casting, high basicity ($1.2 \leq (\%CaO)/(\%SiO_2) \leq 1.4$) mold fluxes are widely used to ease the crack tendency of slabs. The mechanisms by which such mold fluxes control heat transfer are widely considered to be as follows [3–6]:

(1) High basicity increases the solidification temperature (break temperature) of liquid slag, which increases the thickness of solid slag film between mold and steel shell for heat flux control.
(2) The crystallization in the glass layer of solid slag film (devitrification) increases the roughness of film surfaces in contact with the copper mold, which increases the thermal contact resistance between mold and slag film.
(3) Compared with low basicity mold fluxes, the crystals precipitated by solidification in high basicity slag films decrease the thermal conductivity of solid slag film.

Although high-basicity mold fluxes decrease the cracking tendency of peritectic steel slabs, their use may cause lubrication problems like sticking or even breakouts (suggested to be related to the thicker solid and thinner liquid slag film [3, 4]). The most commonly used method to resolve this contradiction (between heat transfer control and lubrication) is to decrease the casting speed. To provide more information to solve this problem, an improved water-cooled copper probe was used to solidify slag films from a typical traditional high-basicity flux (TB), using different probe immersion times. Ultrahigh-basicity mold fluxes developed by Qian Wang et al. of Chongqing University [7–9] to overcome the limitations of traditional high-basicity films were also tested.

Experimental Method

Slag Film Formation

Compositions of the traditional high-basicity mold flux (TB, based on current industrial practice) and the ultrahigh-basicity mold flux (HB) used in this work are summarized in Table 1. Fluxes were prepared using analytical reagents (CaF_2, $CaCO_3$, SiO_2, Na_2CO_3, Li_2CO_3, MgO, and Al_2O_3). To obtain solid slag films, 300 g of slag sample was placed in a graphite crucible with 60 mm inner diameter and melted at 1400 °C in a resistance furnace. The slag film was solidified on a

Table 1 Composition of mold fluxes (wt%)

Slag	%CaO/%SiO_2	F	Al_2O_3	MgO	$Li_2O + Na_2O$
HB	1.74	9.5	4	3	10.25
TB	1.38	9.8	3.83	1.33	12.04

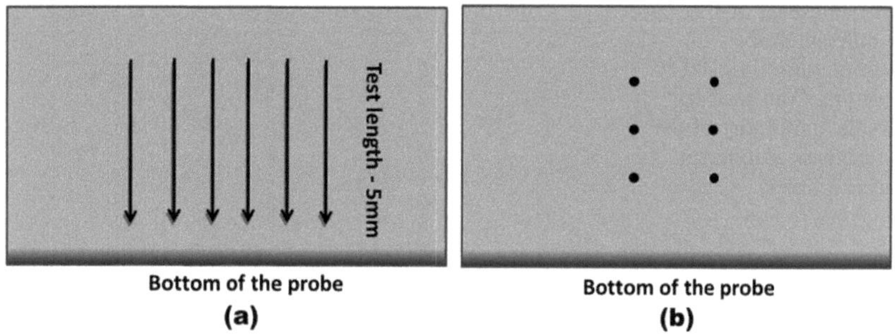

Fig. 1 The direction and position of surface roughness measurements **a** and the positions of thickness measurements **b** of slag films recovered from the wide face of the water-cooled probe

water-cooled copper probe (water flow rate: 1.7L/min) with a smaller size (Width 20 mm × Height 15 mm × Thickness 6.35 mm) and larger width-thick ratio than previously used probes [10, 11]. To reveal the solidification process of films, different probe immersion times (15, 30, 45, and 60 s) were used.

Slag Film Analysis

The surface roughness (of the surface which had been in contact with the probe) was measured with a contact profilometer and the film thickness was measured with a point micrometer, using six measurements of each for every film. The specific positions of the measurements are shown in Fig. 1. The internal structure of the film was inspected by optical and scanning electron microscopy (SEM) after mounting samples in resin, polishing and sputter coating with 2 nm of platinum for SEM. The overall closed porosity of film was measured by the density difference between solid films and films pulverized to an average powder particle size smaller than 45 μm. The density of solid and pulverized films was measured by an Ultrapycnometer 1000.

Results and Discussion

Thickness and Growth Rate of Solid Films

The thicknesses of traditional and ultrahigh-basicity solid slag films after different probe immersion times are shown in Fig. 2, and the corresponding average growth rates for the different time periods are shown in Fig. 3. As expected, film thickness does not increase linearly with time. For the TB flux, the average growth rate of solid films from 0 to 15 s of probe immersion is relatively high (approximately

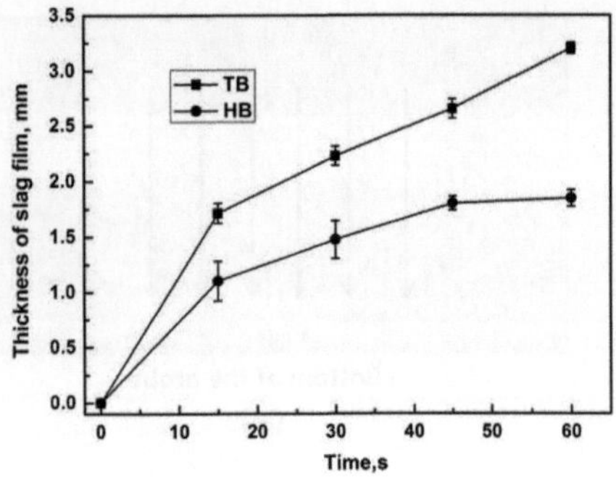

Fig. 2 *Thickness* of slag film after different probe immersion times in 1400 °C molten flux (The error bar shows the uniformity of the film thickness at different immersion times)

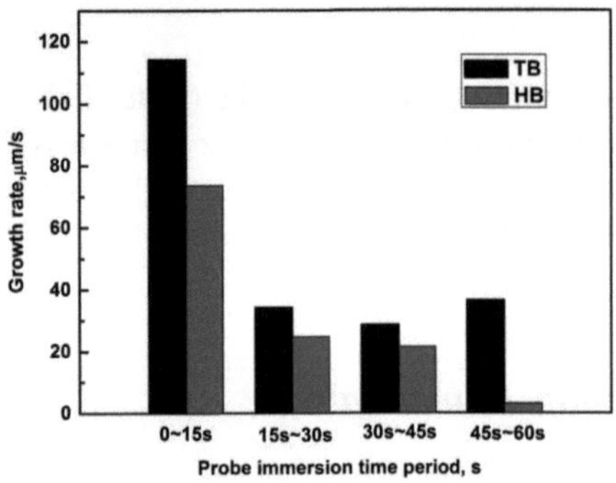

Fig. 3 Average growth rates of solid slag films for different probe immersion periods

110 μm/s), but lower (around 30 μm/s) for longer times, up to 60 s. For the HB flux, the average growth rate of films gradually decreases from approximately 70 μm/s (0–15 s) to approximately 3 μm/s (45–60 s). The thickness of films become stable after the probe immersion time exceeds 45 s. If a thick film (as observed for the TB flux) was to form under mold conditions, the expected decrease in the thickness and uniformity of the liquid films might lead to lubrication difficulties.

Surface Roughness of Solid Films

The roughness of film surfaces (in contacted with the mold wall) increases the thermal contact resistance between mold and steel shell, which plays a significant role in the heat transfer rate under caster conditions. The measured surface roughness of films obtained after different probe immersion times is shown in Fig. 4. The results show that the different in surface roughness of the films is not statistically significant. HB films recovered after longer immersion times did show statistically significant smaller surface roughness, but not so in the case of the TB films.

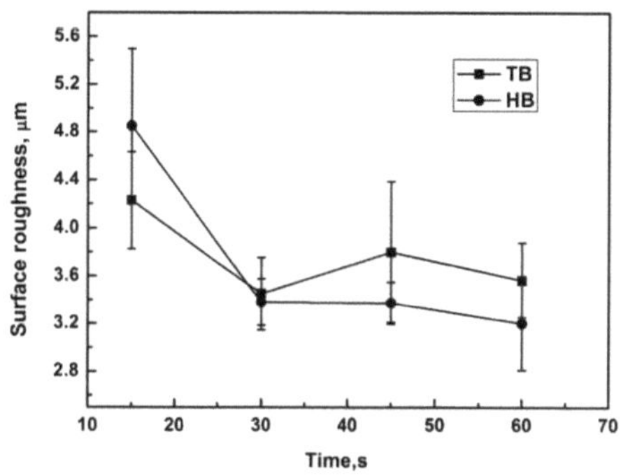

Fig. 4 Roughness of the slag film surface which had been in contact the probe, for different probe immersion times in the molten flux (The error bar shows the uniformity of the surface roughness of films at different immersion times)

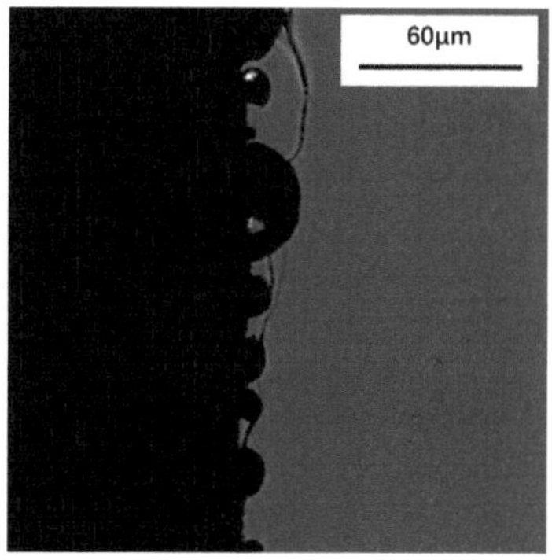

Fig. 5 Typical micrographs of the section of films, showing the surface which had been in contact with the probe (backscattered electron image)

The observation that the roughness does not increase with increased probe immersion times (and even decreases in HB films) is significant, in that it demonstrates that surface roughness is not caused by the precipitation of crystals in the glassy layer of solid films; the same (or higher) surface roughness was found for mostly or fully glassy films formed after short immersion times. The surface features visible in Fig. 5 indicate that pores formed at the film-probe interface contribute to the formation and evolution of roughness.

In the continuous casting of peritectic steels, the initial steel shell formed at the meniscus is very weak and sensitive to non-uniform cooling which promotes the formation of cracks. A large surface roughness of solid mold flux films is expected to ameliorate fast and non-uniform cooling of the steel shell at the meniscus.

Measured Porosity of Solid Films

The presence of pores in solid film can affect film conductivity. A density method was developed to measure the overall closed porosity in films. The porosity of slag film can be calculated as follows:

$$P = \frac{(D_P - D_S)}{D_P} \times 100\%$$

P—porosity of slag film;
D_p—true density of pulverized slag film;
D_s—density of solid slag film.

The measured porosities of film samples (Fig. 6) show that the overall porosity of HB films is much higher than the TB films. The resulting lower effective thermal conductivity may have contributed to the thickness of these films. Cross-sections of the films (Fig. 7) show that in the TB films most of the pores formed close to the probe, thus the overall porosity decreased with increased film thickness. For HB films, the porosity increases with increased probe immersion times. The large lath-shaped crystals formed during solidification are associated with the large porosity in the crystalline layer of these films.

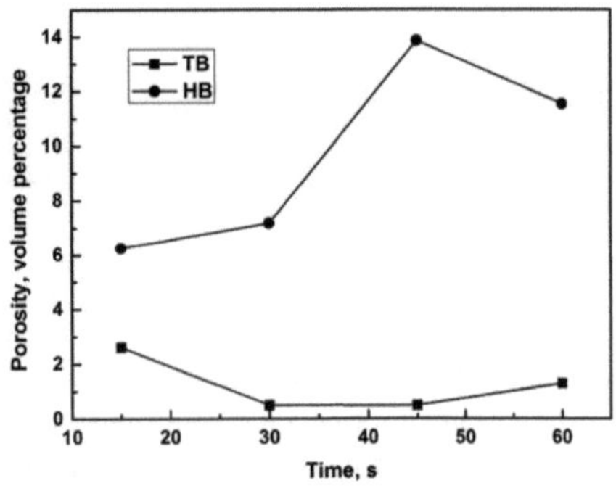

Fig. 6 The porosity of slag film with different probe immersion times

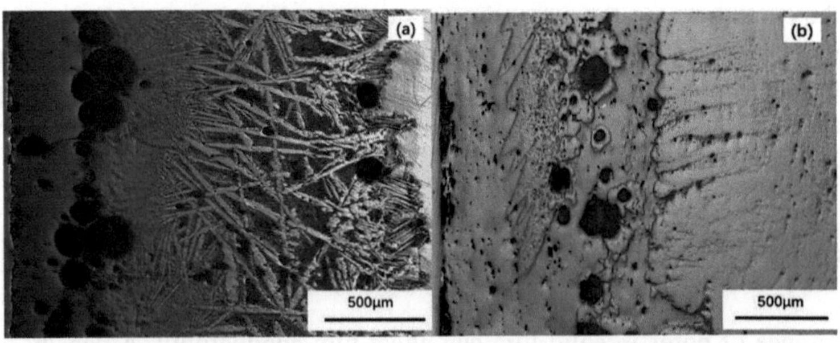

Fig. 7 The section features of HB **a** and TB **b** slag films (optical micrographs)

Table 2 quality statistic of slabs

Year	Total production, millions of tons	Longitudinal crack ratio, %	Scrap ratio, % (caused by Longitudinal cracks)
2012	5.29	5.32	0.13
2013	5.38	4.92	0.028
2014	3.82	2.15	0.00061

Table 3 Defect ratio (%) of final products caused by longitudinal surface cracks

Year	Section width of heavy and medium plates		
	4100 mm	2700 mm	1780 mm
2013	0.43	0.09	0.417
2014	0.14	0.03	0.065

Industrial Applications

In Chongqing Iron & Steel Co. China, heavy and medium plates (4100, 2700 and 1780 mm in width) from hypo-peritectic steel slabs are major products. According to reported statistics [12], the longitudinal cracking ratio of hypo-peritectic steel slabs in the No. 1 steelmaking plant from late 2009 to November 2012 ranged from 1 to 30%, in the worst case reaching 1000 tons of steel scrap per month caused by longitudinal cracks. Based on the steel grades and production procedure, a series of ultrahigh-basicity mold fluxes (basicity 1.7–1.8) was developed by Chongqing University [7–9] and applied on three casters in the No. 1 steelmaking plant from November 2012. Two-year statistics from 2012 to the end of 2014 indicates that after starting to use ultrahigh-basicity mold fluxes, the longitudinal cracking ratio and especially the scrap ratio caused by large cracks decreased significantly. The subsurface crack ratio also decreased significantly (see Tables 2 and 3). The data indicate that compared with traditional fluxes, the ultrahigh-basicity mold fluxes have good performance in improving the surface quality of slabs and final products. Over a period of two-years (2013 and 2014), no exceptional conditions such as sticker breakouts occurred.

Conclusions

Based on the comparison between the selected traditional basicity mold flux (TB slag) and ultrahigh-basicity mold flux (HB slag), and the large-scale application in industrial processes, some conclusion can be drawn:

(1) Compared with TB slag films, the thickness and growth speed of HB slag film are smaller, which is expected to improve the lubrication capacity of mold fluxes.
(2) Pores contribute to the surface roughness of slag films. HB films have higher roughness at the earlier stage of solidification, which decreases when the probe immersion time exceeds 15 s. For TB films, the effect of probe immersion time on roughness is not obvious.
(3) The overall porosity of HB films is higher than TB films have.
(4) Based on two years of industrial use, the HB slag improved the surface quality of slabs. No exceptional conditions like sticking breakout occurred.

Acknowledgements The authors greatly appreciate the support from the industrial members of the Center for Iron and Steelmaking Research at Carnegie Mellon University, Chongqing Graduate Student Research Innovation Project (CYB14035), China Postdoctoral Science Foundation (2016M592119) and Jiangxi Postdoctoral Research Project (2016KY19).

References

1. K. C. Mills, A. B. Fox, "The Role of Mould fluxes in continuous casting-so simple yet so complex." *ISIJ International*, 43(10) (2003), 1479.
2. H. Nakada et al., "Mechanism of heat transfer reduction by crystallization of mold flux for continuous casting." *ISIJ International*. 48 (4) (2008), 446.
3. J. Chi and Y. Gan, *Mold fluxes*. (Shen yang, China: Northeast University Press, 1992), 5–25.
4. S. Seetharaman, ed., *Treatise on process metallurgy* (Oxford: Elsevier, 2013), 435–475.
5. K. Tsutsumi, T. Nagasaka and M. Hino. "Surface roughness of solidified mold flux in continuous casting process." *ISIJ International*, 39(11) (1999), 1150.
6. M. Susa et al., "Effects of both crystallisation and iron oxides on the radiative heat transfer in mould fluxes." *ISIJ International*, 49 (11) (2009), 1722.
7. X. Long et al., "Properties of High Basicity Mold Fluxes for Peritectic Steel Slab Casting." *Journal of Iron and Steel Research International*, 19(7) (2012), 39–45.
8. X. Long et al., "Effects of Crystallization of Mould Fluxes on Property of Liquid Slag Film and Its Impact on Peritectic Steel Slab Continuous Casting" in *4th International Symposium on High-Temperature Metallurgical Processing*, (Warrendale, PA: TMS, 2013), 155–162.
9. S. He et al., "Effects of crystallisation behaviour of mould fluxes on properties of liquid slag film" *Ironmaking & Steelmaking*, 39(2012), 593–598.
10. K. Lara Santos Assis, "Heat transfer through mold fluxes: a new approach to measure thermal properties of slags" (Ph. D. Thesis, Carnegie Mellon University, Pittsburgh, 2016).
11. H. Ryu et al., "Crystallization behaviors of slags through a heat flux simulator." *ISIJ International*. 50(8) (2010), 1142.
12. Y. M. He et al., "Application of high-basicity mould fluxes for continuous casting of large steel slabs" *Ironmaking & Steelmaking*, 43(8) (2016), 588–593.

Part II
Metal Extraction

Applied Statistical Analysis on the Calcination Process in the Ferronickel Production

Fabio M. Soares, Denis Shevchenko, Alexey Levchenko, Alexey Avdeev, Alexander Vodin, Vitaly Rudik and Stanislav Kovalchuk

Abstract Ferronickel is mainly produced by the RKEF (Rotary Kiln Electric Furnace) process. The ore is extracted, crushed and dried before calcination. Rotary kilns are usually about 100 m long and rotate to facilitate material flow. Due to this length, in addition to rotation speed, the material takes a variable time to cross the whole kiln, and thereby changing the chemical and temperature profile, making the control of calcine temperature a great challenge. However, statistical analysis are a great tool for finding interesting patterns that are of valuable help to control the kiln variables, that are very sensitive to inertia caused by rotation and changes in temperature profile. We present a study based on real data taken from a processing plant, whereby we applied data mining techniques to extract information on which variables have influence on kiln's key performance index variables, such as calcine temperature.

Keywords Rotary kiln control · Applied statistics · Chemical process

Introduction

Statistics has been applied on process control for decades by means of techniques like Analysis of Variance (ANOVA), Statistical Process Control (SPC), Principal Components Analysis (PCA) and Projection to Latent Structures (PLS) [1]. These techniques offer great help to control very sensitive process variables.

Nowadays every industry maintains huge process databases that implicit reflect the process dynamics, whereby these data provide the underlying base for all kinds of process analysis. In this sense, it seems reasonable to look back at those data to examine some unexpected deviation, or investigate some process phenomena [2]. In addition, the history database hides invisible patterns, even for experts in operation,

F.M. Soares (✉) · D. Shevchenko · A. Levchenko · A. Avdeev · A. Vodin · V. Rudik · S. Kovalchuk
Federal University of Pará, Belém, Brazil
e-mail: fabiosoares22@gmail.com

A. Allanore et al. (eds.), *Materials Processing Fundamentals 2017*,
The Minerals, Metals & Materials Series, DOI 10.1007/978-3-319-51580-9_2

but detectable by statistics [3]. Understanding these patterns is fundamental for carrying out a great control.

In this work, we present a statistical procedure to find patterns in the calcination phase of the Ferronickel RKEF (Rotary Kiln Electric Furnace) process.

The Rotary Kiln in the Ferronickel Production

One of the ferronickel production forms is by the RKEF process. In this process, run-of-mine ore is extracted, crushed, dried before being fed into rotary kilns [4]. The main role of rotary kilns is to transform chemically the ore into calcine by heating it up to around 900 °C over the time needed to calcination process. The rotary kilns for this process are usually long, extending over 100 m, leading the material to take hours to cross the entire kiln, therefore resulting in a temperature gradient over the kiln and three defined zones: drying, heating and reduction zones (Fig. 1).

Other processes, such as cement and lime calcining, also apply rotary kilns [5]. The length should be compatible to a proper heating and reduction time, wherein rotation acts as a significant component to the material passage. A burner is located at a lower level to provide heat throughout the kiln, and gas exhausters account for the flame extension and a smother temperature gradient. An exhauster pulls out the hot gas through a bag filter, cleaning it before throwing into atmosphere. This exhauster helps also in the flame length due to the pressure drop. A blower injects additional (secondary) air into the kiln, to help combustion. Figure 2 shows a schema of the whole process.

The Calcine Temperature Challenge

In every rotary kiln application (not only for ferronickel production), calcine temperature is the most important KPI (key performance index). Insufficient calcine temperatures lead to bad calcine quality [6, 7]. On the other hand, a very high temperature can cause problems to the refractory lining at the hot end [4]. For those reasons, there is an optimum range for the calcine temperature. However, it is influenced by many factors in the kiln, including:

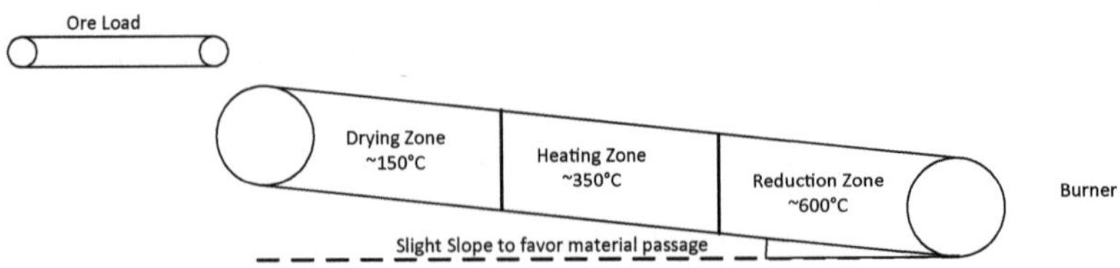

Fig. 1 Kiln scheme with three zones

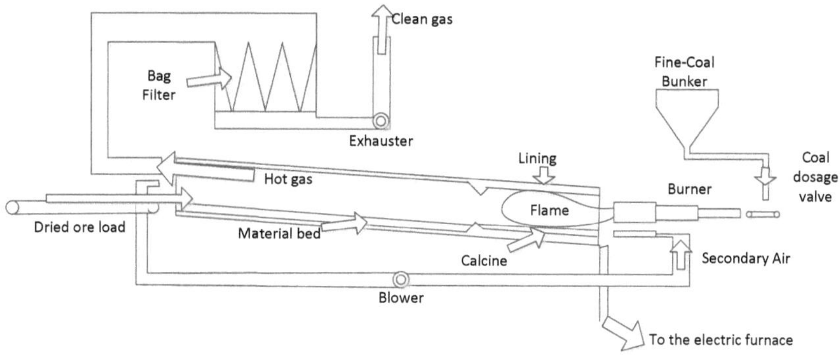

Fig. 2 The rotary kiln full system

- Pressure drop
- Temperature profile
- Chemical profile
- Kiln rotation
- Primary, secondary and tertiary air flows
- Material inflow and outflow rate.

The kiln rotation is tightly linked with the material flow. Kiln should rotate faster as ore load increases. However changes in rotation cause changes in temperature profile. Thereby calcine temperature responds to control actions only after a variable time. This fact makes both the modelling and control of the rotary kiln very complex tasks [8, 9]. Current approaches are just approximations of the time-varying plant. However, artificial intelligence and data mining based models are able to give good results [5].

Data Mining and Statistical Analysis in Industrial Processes

Data mining and statistical analysis techniques are helpful in finding interesting patterns in the plant operation [3]. As the technology evolved to collected millions of data over a very short time, the analyses of such data are still in development though [2]. Computer control does not rely only on the classical digital control techniques anymore, but also on the tons of data produced by huge sensor networks. This calls for the data mining techniques to find useful information in this huge dataset.

Pattern Recognition

In industrial processes, there are a number of data mining applications to formally define the process patterns. Wang et al. [10] exploited data mining to predict, optimize and diagnose parameters in oil field. Namikka and Gibbon [2] applied data mining for exploratory analysis and process modelling making use of multivariate statistical methods for dimensionality reduction. Meré et al. [11] reviewed the typical environments where data mining is applied, and presented a series of tools to deal with outliers in the galvanized steel production. Sadoyan et al. [12] sought patterns in manufacturing process control using Rough Sets (RS) for identifying if-then type rules.

Regarding process control, artificial intelligence techniques like Fuzzy Systems and Neural Networks are used in the works of Yang et al. [9] and Meré et al. [11], however these techniques have a complex implementation and require large computational time [12]. In addition, neural networks are black box, i.e. one does not know how the results are produced [8]. Techniques like Association Rules [13] and Decision Trees [14] are able to find relevant patterns in the process in a clearly understandable text form.

Control Variables Histogram

Considering the process as an N-dimensional space, rules define a subspace wherein each variable will have a new statistical distribution, in a histogram form. By classifying the variables into control (U) and observation (Y), and assuming that the control will reflect after some time on the observation, we can apply joint probability distribution (JPD) to statistically determine the relation between control and observation:

$$P(Y,U) = P(Y)P(Y|U) \tag{1}$$

Where *P(Y)* is a probability for the vector of observations *Y* to have a certain set of values, *P(Y|U)* is the conditional probability of the observations *Y* to have a set of values given that *U* has a set of control values. The conditional probability can be interpreted as the value of *Y* depends or is influenced by the value of *U*. In a complex system, this may be modelled as a Bayesian Network, which is a diagram of causation theoretically design by an expert in the field of application. Deventer et al. [15] show an example of a Bayesian Network applied in control of dynamic process (Fig. 3).

The probability distributions (PD) are determined based on the process history in the form of histograms. By applying one rule (filter or slice in the n-dimensional

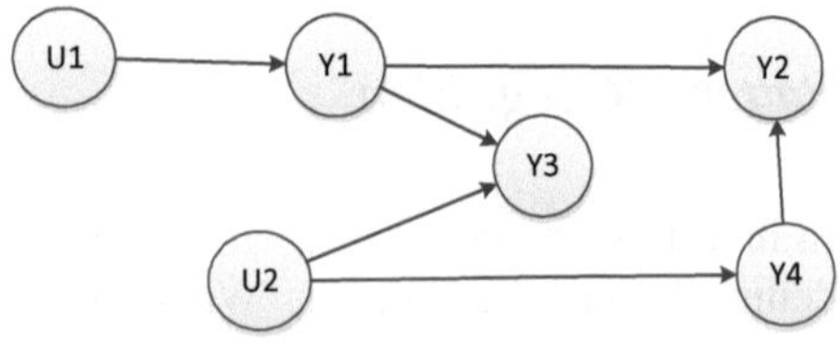

Fig. 3 Schema of a multivariate Bayesian Network with 2 controls and 4 observations

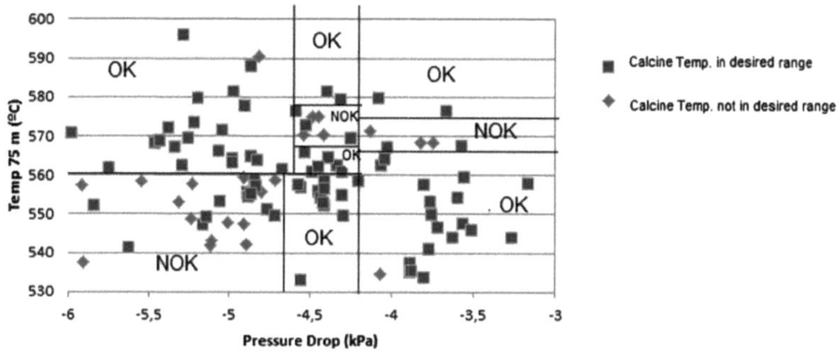

Fig. 4 Scatter plot of two variables where the desired range points are separated by slicing the dataset

Fig. 5 A decision tree that separates the desired from the undesired range in the previous dataset, considering variables Pressure Drop (PD) and Temperature at 75 m (T75)

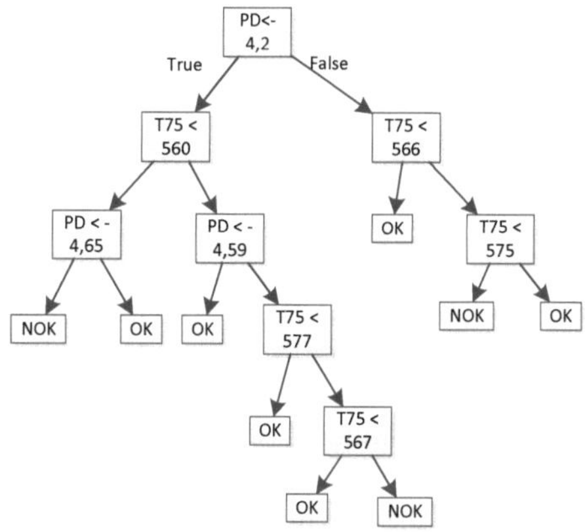

data hyperspace), the PD’s change until we drill down to a single record containing only one single value for each variable.

Determining Optimal Control Ranges

For the main performance index, we define the range on which it should be. That would be our target parameter Y_p, and $P(Y_p = [y_{pmin}\ y_{pmax}])$. So, we need to find the slices in the data for the control and observable variables that would maximize $P(Y_p = [y_{pmin}\ y_{pmax}] | slices)$. These slices will compose a certain rule of the type *“if* $Y = [y_{min}\ y_{man}]$ *and* $U = [u_{min}\ u_{max}]$ *then* $Y_p = [y_{pmin}\ y_{pmax}]$ *with probability P”*.

Since we are interested only on the desired performance index Y_p, we can leave only the desirable values, reducing the rule to the form *“if* $Y = [y_{min}\ y_{max}]$ *then* $U = [u_{min}\ u_{max}]$ *with a probability P”*. The rules can be determined using information criteria applied via association rules and decision trees techniques (Figs. 4 and 5).

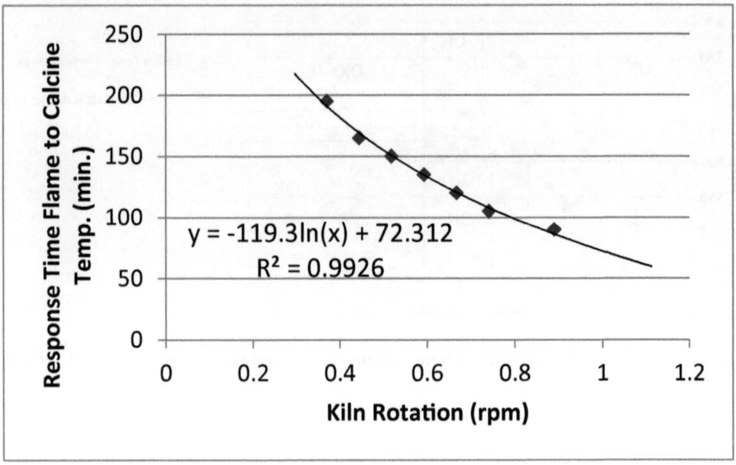

Fig. 6 Response time of Flame to Calcine Temperature according to Saeman [16]

Dynamic Part of the System

To handle the delays of the system, we applied two approaches. Firstly, we used the Saeman [16] residence time equation to determine proportionally the response time to control according to kiln's rotation.

That resulted in the following relation with the response time (Fig. 6).

Another additional strategy to determine time differences is by tracking the influences via the theoretical causal relations. It is known by the work of Naud & Emond [7], that the fuel combustion causes the flame to heat up the lining, which is a refractory material, i.e. it retains the heat. The material then receives the heat from the lining, besides the flame itself. So the more accurate way to investigate the temporal response is by determining the influence of the fuel valve on a temperature point nearest to the flame's highest temperature, and then the influence of this point to the calcine's temperature. In our case, this point is 75 m from the cold end (Figs. 7 and 8).

By performing correlation analyses and filtering out noisy data, we found the following relation.

Fig. 7 Dependency relation between Fuel, Material and Calcine Temperature

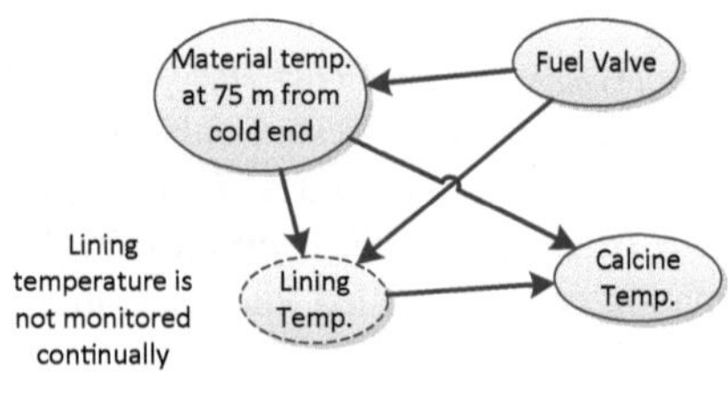

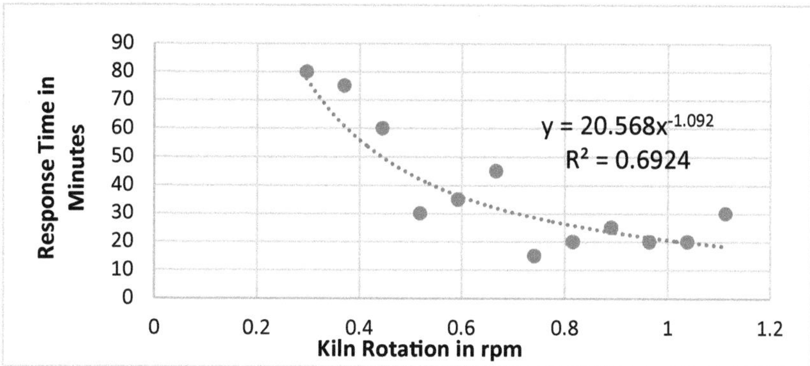

8a – Response time of calcine temp. to temp. 75m

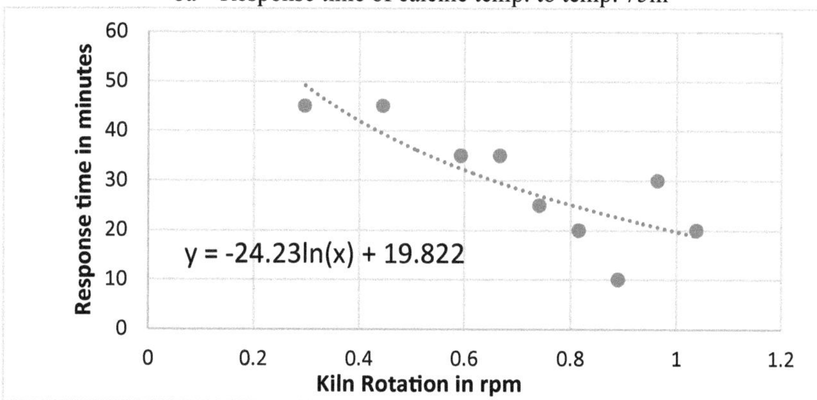

8b – Response time of flame to temp. 75 m.

Fig. 8 **a** Response time of calcine temp. to temp. 75 m **b**– Response time of flame to temp. 75 m

Finally, to find the rules, we collected 4 months (April thru August 2016) of data from our process plant and sliced up this dataset into 10 zones corresponding to 10 rotation speeds (from 50 T/h to 140 T/h). The total size of the database was about 152000 records sampled by minute. The variables included in the control were:

- Ore Load
- Temperature at 48 and 75 m from cold end
- Oxygen concentration
- Pressure drop before exhausters
- Secondary air pressure
- Calcine Temperature
- Fuel load.

Comparison with Current Control

The realization of a control algorithm involves a number of practices including programming and integration with production environment [5]. This step can be unexpectedly long on account of technical resources. In every case, it is possible to make comparisons with the current control used and somehow determine if the new strategy is able to detect or prevent undesirable deviations in the process. At the time of writing this article, the control algorithm was not yet implanted in production environment, but we performed a simulation of what control actions would be taken for every minute of operation.

In Fig. 9 we plotted the results of a comparison over one work day in the plant. The light yellow line with circles at the top refers to the calcine temperature; the dark line with triangles is the actual control actions taken, and the dark blue line with circles at the bottom refers to the ore load. The gray lines represent optimal the interval for the control that was determined via the rules found by the decision tree or general association rules algorithm.

Some interesting and worth noting cases are discussed as follows. During the preheating phase, the lining may heat up very quickly while the material that remained in the kiln takes longer time to heat, causing sudden variations in temperature profile. Moreover, during preheat the ore load and rotation is low, leading to a long response time of the calcine temperature, and because of that, we take into account only ore loads greater than 50 T/h. In this figure, although the lining heated up quickly, the material that came afterwards was not sufficiently heated, as the kiln rotated faster. The algorithm suggested for that state (ore load around 50 T/h) one extra ton for coal dosing (3.5 T/h instead of 2.5 T/h).

The numbers in the figure indicates some other cases:

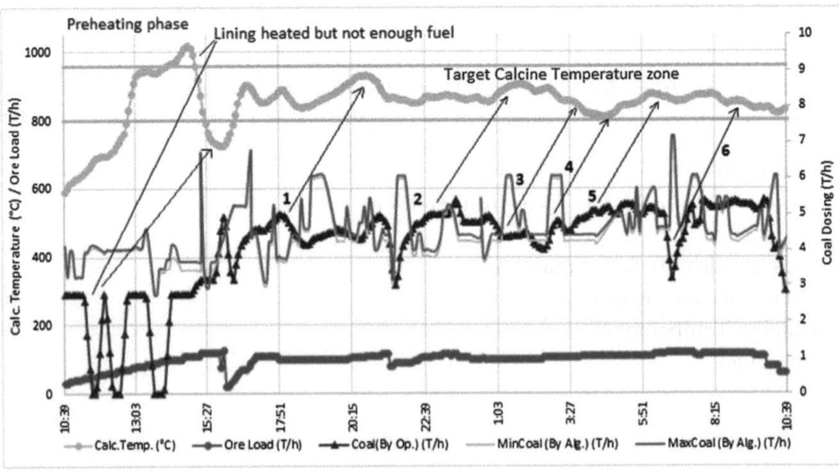

Fig. 9 Comparison of this data analysis approach with current control

1. The calcine temperature was in a rise trend and the algorithm suggested a lower level of fuel
2. Same case of 1
3. Forecasting a fall in the calcine temperature, suggested an additional fuel load
4. Although the calc. temperature was falling, the algorithm caught signs that the temperature was about to rise again, and suggested a lower level of fuel after having suggested a higher level
5. To prevent temperature rise, suggested a lower level
6. Some event may have cause calcine temperature to fall in the next hours and the control suggested a high load of fuel.

It is worth noting that this simulation should be dynamic, i.e. if the control actions taken by operators were the same of the suggested ones, the next suggestions would surely be different, provided that the observable variables would have changed. This may explain sudden and abrupt variations in the coal dosing suggestions. Since this algorithm is purely based on statistics, it is very likely that the kiln eventually matches a rule, which would suggest a very different coal dosing level than the previous or current level. Fuzzy decision trees [17] would certainly prevent such sudden changes in suggestions, as the bounds between rules would be soft and gradual instead of crisp hard limiting. It should also be considered that this algorithm is based on real data, events that actually happened in the plant, so new events that can affect calcine temperature are not detected. In any case, actions that an experienced operator or an expert in kilns would take are likely to be represented by these rules; therefore, new staff will have access to this knowledge by means of this system.

Conclusions

In this paper we have showed a statistically based approach to find patterns in the calcination process in ferronickel production. Since this is a very complex and hard to control environment, statistics arise as a powerful tool to discover relations between events and variables towards a satisfactory end. In our case the task was to find the fuel levels that would provide the desirable range in calcine temperature with the maximum likelihood. By applying data-based algorithms, like association rules and decision trees, we found the searched patterns in the form of rules, whereby the optimum control ranges are determined by probability.

Although this control is not applied to production yet, we could get a glimpse of some results by static simulation with real time process data. The results found in this article will support the adoption of this methodology to search for new patterns in this process.

References

1. Semel, F. J., D. J. Kasputis, and P. U. Gummeson. "Statistical Process Control in Iron Ponder Production and new product development." *SAE Congress.* Detroit, Michigan, 1988.
2. Namikka, E., and G. J. Gibbon. "Identification of data mining techniques for industrial process analysis and control." *1st African Control Conference AFCON.* Cape Town, 2003.
3. Platon, R., and M. Amazouz. *Applicaton of Data Mining Techniques for Industrial Process Optimization.* Natural Resources of Canada, 2007.
4. Shaobao, Li. "Study of Nickeliferrous Laterite Reduction." Hamilton: McMaster University, 1999.
5. Gao, X., and Li Zhang. "Survey on rotary kiln process control." *Chinese Control and Decision Conference (CCDC).* 2009.
6. Shatilov, O., A. Chelpanov, и S. Chuykov. (Automatized rotary kilns regulator and control system in Russian), 3 2002 г.: 20–27.
7. Naud, Serge, and Martin Emond. "Lime Kiln Control using Simple Advanced Regulatory Control Strategy." *ISA EXPO.* Houston, TX, 2007.
8. Ginsberg, Tobias. "Dynamische Modellierung von Drehrohröfen." (Dynamic Modelling of Rotary Kilns - in German) Aachen: Rheinisch-Westfälische Technische Hochschule Aachen, 2010.
9. Yang, B., X. Ma, and Q. Zhang. "Elman neural network based temperature prediction in cement rotary kiln calcining process." *Intelligent Systems and Knowledge Engineering ISKE.* 2010.
10. Wang, Tong, Xian-wen Gao, and Kun Li. "Application of data mining to production operation and control system in oil field." *24th Chinese Control and Decision Conference (CCDC).* 2012.
11. Meré, J. B.O., M. C. Limas, F.J. M.P. Ascacíbar, and A. G. Marcos. "Data Mining in Industrial Processes." *Congreso Español de Informática CEDI.* Granada, Spain, 2005.
12. Sadoyan, H., A. Zakarian, and P. Mohanty. "Data mining algorithm for manufacturing process control." *International Journal of Advanced Manufacturing Technologies* 28 (2005): 342–350.
13. Pang-Ning, T., M. Steinbach, and V. Kumar. "Classification: Basic concepts, Decision Trees and model evaluation." In *Introduction to Data Mining*, 145–205. Addison Wesley, 2006.
14. Jung, Y. K., Lee, J. "Multivariate process control procedure using a decision tree learning technique." (In Korean) *Korean Data and Information Society*, 2015: 639–652.
15. Deventer, R., J. Denzler, and H. Niemann. "Control of Dynamic Systems Using Bayesian Networs." *Proceedings of IBERAMIA/SBIA 2000 workshops.* Atibaia, São Paulo, 2000.
16. Saeman, W. C. "Passage of solids through rotary kilns: factors affecting time of passage." *Chemical Engineering Program 47,* 1951: 508–514.
17. Olaru, Cristina & Wehenkel, Louis, A complete fuzzy decision tree technique, *Fuzzy sets and systems 138*, 2003: 221–254.

The Chemical Stability and Electrochemical Behavior of Dy_2O_3 in Molten $CaCl_2$

Jianxun Song, Bart Blanpain and Jan Fransaer

Abstract The chemical stability of Dy_2O_3 in molten $CaCl_2$ was studied by confocal scanning laser microscopy. Results illustrate that Dy_2O_3 is relatively stable in molten $CaCl_2$, and it is a good candidate for the extraction of dysprosium or dysprosium alloys through direct electrochemical reduction. The electrochemical behavior of dysprosium oxide in contact with either a solid molybdenum or a liquid tin cathode has been investigated by cyclic voltammetry in a $CaCl_2$ melt at 1173 K (900 °C).

Keywords Dy_2O_3 · Molten $CaCl_2$ · Direct reduction · CSLM · Liquid electrode

Introduction

Dysprosium is produced by calciothermic reduction of DyF_3, reduction distillation of Dy_2O_3 followed by master alloy distillation of Dy–Mg and electrochemical reduction of Dy from $DyCl_3$ in molten salts [1–3].

Compared with calciothermic or magnesiothermic reduction method, molten salt electrolysis is widely used to prepare RE metals and alloys because of its simplicity and high processing rate. The cathodic electrochemical reduction of solid oxides, which is called the Fray-Farthing-Chen (FFC) Cambridge process, has been studied for the preparation of titanium from TiO_2 since 2000 [4]. Actually, the method could be applied to produce many other metals, such as Nb, Ta, Zr, Cr, W and rare earth metals. Kim et al. [5] and Xie et al. [6] studied the direct electrochemical reduction of Dy_2O_3 in $CaCl_2$ melts. It was reported that Dy_2O_3 could be reduced to metallic dysprosium and it is one step process from Dy(III) to Dy. So far, no other paper was published on the dysprosium extraction from Dy_2O_3 especially on a liquid electrode.

J. Song (✉) · B. Blanpain · J. Fransaer
Department of Materials Engineering, KU Leuven, Kasteelpark Arenberg 44, 3001 Heverlee, Belgium
e-mail: Jianxun.song@kuleuven.be; jensonsoong@gmail.com

A. Allanore et al. (eds.), *Materials Processing Fundamentals 2017*,
The Minerals, Metals & Materials Series, DOI 10.1007/978-3-319-51580-9_3

Thus, in order to optimize the direct reduction process, the chemical stability and the electrochemistry of Dy_2O_3 in molten $CaCl_2$ at 1173 K (900 °C) was studied using confocal scanning laser microscopy and cyclic voltammetry respectively.

Experimental Methods and Materials

Sintering of Dy_2O_3 and quenching experiments

Dy_2O_3 powder (Chempur, 99.99%) was molded into a disk (10 mm diameter, 5 mm thickness) and pressed with a tablet press (Paul-Otto Weber D7064) at 30 MPa. A cold isostatic press (CIP, EPSI B-9140) was used at a pressure of 2000 bar with a dwelling time of 5 min. The samples were then heated with a heating rate of 3 K min^{-1} and sintered at 1173 K (900 °C) for 12 h. It was cooled down to room temperature at a cooling rate of 3 K min^{-1}. The temperature was measured using a calibrated thermocouple placed directly above the sample. The microstructure of the sintered Dy_2O_3 was investigated by scanning electron microscopy (SEM, Philips XL30 FEG).

In the quenching experiments, $CaCl_2$ (Aldrich, 99.9%) electrolyte was pre-dried under vacuum for 24 h at 573 K (300 °C). Subsequently, it was melted at 1173 K (900 °C) in a highly pure argon atmosphere (99.999%). The pre-electrolysis process was performed between a Kanthal wire cathode and a graphite anode at a potential of 2.5 V during 2 h to remove moisture and impurities. The $CaCl_2$ salt was remelted in a quartz cup (25 mm diameter, 25 mm height) under Ar gas and a sintered particle of Dy_2O_3 (6 mm diameter, 2 mm thickness) was dropped into the crucible after the temperature reached 1173 K (900 °C), after which the quartz cup was held at that temperature for 2 h and then quenched in air.

The cold mounted samples were polished with 1 μm diamond paste after grinding with emery paper 80, 320, 600, 800, 1200 and 1400, respectively. The as-polished samples were then platinum coated and analyzed by SEM. Qualitative element maps of calcium, dysprosium, chlorine and oxygen were recorded to determine the composition in the bulk and at the oxide/salt interface.

The dissolution behaviors of Dy_2O_3 in molten $CaCl_2$

The spherical Dy_2O_3 particles used in CSLM-IIF observation were prepared by grinding sintered particles of Dy_2O_3 of approximately 1 mm. The diameter of the final spherical Dy_2O_3 particles is 350–400 μm.

In-situ observations of the dissolution of Dy_2O_3 in molten $CaCl_2$ were carried out with a resolution of 0.25 μm using the CSLM-IIF equipment. Details of the CSLM technique have been published elsewhere [7–10]. A recrystallized alumina crucible (5 mm diameter and a height of 8 mm) was used in this experiment, and it was placed in a bigger platinum crucible (10 mm diameter and a height of 10 mm) against accidental spillage. The $CaCl_2$ employed in the CSLM experiments was prepared using the above procedures. The pre-melted $CaCl_2$ was loaded into the crucible, heated in ultra-high purity argon (99.999%) in the CSLM up to 1173 K (900 °C) and cooled to form a uniform flat surface. The spherical Dy_2O_3 particles were placed on the surface of the solid salt. The entire assembly was heated to 1173 K (900 °C) under ultra-high purity argon. The details of this process have been described elsewhere [9, 10].

Electrochemical setup and procedure

A three electrode cell assembly was used for all the electrochemical measurements. Two different working electrodes have been used: (1) a metallic cavity electrode (MCE) and (2) a liquid Sn electrode. The MCE was fabricated from Mo foil (2.0 mm width and 1.0 mm thickness, with 1.0 mm diameter circular through-hole cavities), and it was used as a working electrode to record cyclic voltammograms (CVs) [11]. The Dy_2O_3 (99.99%, Chempur) powder was pressed at 15 MPa and sintered at 1173 K (900 °C) for 12 h to obtain a dense pellet which was used as electroactive specie. The sintered material was ground into powder and the MCE cavities were filled by repeatedly pressing this Dy_2O_3 powder on both sides of the foil around the through-holes. Any powder left on the surface of the foil was wiped off carefully. The liquid Sn electrode was prepared by placing Sn shot in a closed-end alumina crucible (ø 6 mm × 6 mm), which was immersed in a $CaCl_2$ melt. A molybdenum wire of 1.5 mm diameter sheathed in a B_4C tube of 5 mm outer diameter was immersed in the Sn phase and used as a lead wire. The surface area of the liquid Sn electrode is approximately 2 10^{-3} m^2. A 6 mm diameter graphite rod served as the counter electrode. A platinum wire (99.99%, 1.0 mm diameter and 30 mm length) was used as pseudo-reference electrode. All the potentials reported in this paper are with respect to this pseudo-reference electrode. Cyclic voltammetry was carried out using a Solartron 1287 potentiostat/galvanostat and the scan rate is 50 mVs^{-1}.

Results and Discussion

The dissolution behavior of Dy_2O_3 in molten $CaCl_2$

Figure 1 shows a typical time series for the dissolution process of Dy_2O_3 in pure molten $CaCl_2$ at 1173 K (900 °C). A circle around each particle was used to enhance the boundary. It can be seen from Fig. 1 that the spherical Dy_2O_3 particle slowly shrinks in time, while the particle remains spherical (even after 7270 s). No surface reaction was observed and the particles remained smooth and spherical during dissolution.

The Dy_2O_3 particle was quenched after it was immersed in molten $CaCl_2$ for two hours at 1173 K (900 °C). The element maps shown in Fig. 2 illustrate that the outline of the Dy_2O_3 particle is well defined. Four elements were detected: Cl, Ca, O and Dy. Figure 2 shows that O and Dy diffused in the melt, but there is no sign of

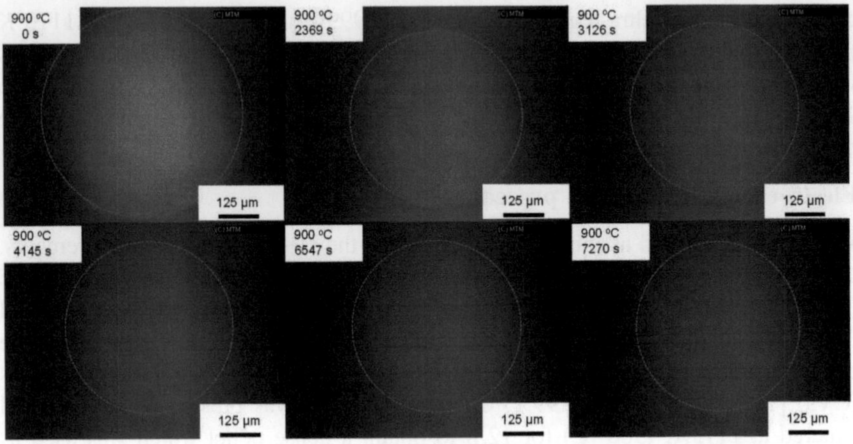

Fig. 1 Time series of CSLM images of a spherical Dy_2O_3 particle in molten $CaCl_2$ at 1173 K (900 °C)

Cl diffusing into the Dy_2O_3. This indicates that no new compounds form on the surface of the Dy_2O_3 particle.

Figure 3 compares the microstructure of the cross section of a Dy_2O_3 particle before (Fig. 3a) and after (Fig. 3b) immersion into molten $CaCl_2$ for 2 h at 1173 K (900 °C). The cross-section was polished with 1 μm diamond paste after grinding with different emery papers. The as-polished samples were then platinum coated and analyzed by SEM. It is can be observed that the particle with a bigger grain size disappeared after it was contact with molten $CaCl_2$(Fig. 3b).

Electrochemical behavior of Dy_2O_3 on a MCE and liquid Sn electrode

The cyclic voltammogram for the MCE electrode in the purified $CaCl_2$ melt is shown by the black dashed line in Fig. 4. It can be seen that there are no electrochemically active impurities in the melt after the pre-electrolysis. The redox wave at E' and the peak E represent the reduction of Ca^{2+} and the oxidation of metallic Ca, respectively. The solid red line in Fig. 4 is the cyclic voltammograms of the MCE with 0.01 wt% Dy_2O_3 loaded between a vertex potential of 0 and −1.6 *vs* Pt. Two pair of redox peaks were observed.

Two corresponding redox peaks are located at −1.38 V and −1.18 V *vs* Pt (B' and B). Two other (less obvious) redox peaks are at −0.85 V and −0.76 V *vs* Pt, labeled A' and A. The cathodic current peak A' is thought to be caused by the reduction Dy(III) to Dy(II). Hence, according to the above analysis, the electrode reactions are:

$$Dy_2O_3 + e^- \leftrightarrow 2DyO + O^{2-} \quad (A/A') \tag{1}$$

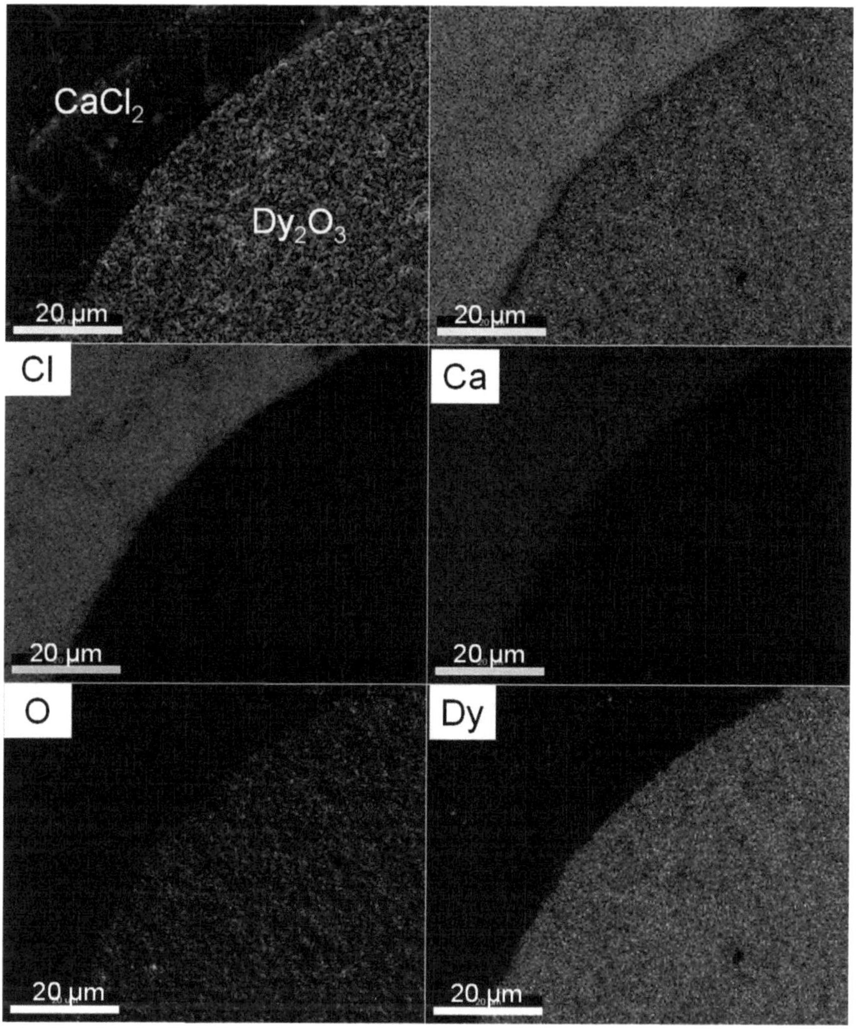

Fig. 2 The EDX mappings of Cl, Ca, O and Dy of Dy_2O_3 after quenching

$$DyO + 2e^- \leftrightarrow Dy + O^{2-} \quad (B/B') \tag{2}$$

Next, a liquid Sn electrode was employed as the working electrode to study the electrochemical behavior of Dy_2O_3 in molten $CaCl_2$.The image shown in Fig. 5 is the liquid tin electrode used in this experiment. Dotted and solid curves in Fig. 5 designate the CVs before and after the addition of Dy_2O_3, respectively. In both cases, the anodic peak at about 0.17 V *vs* Pt corresponds to the dissolution of the Sn electrode. In addition, the current starts to increase at about −0.98 V *vs* Pt in the

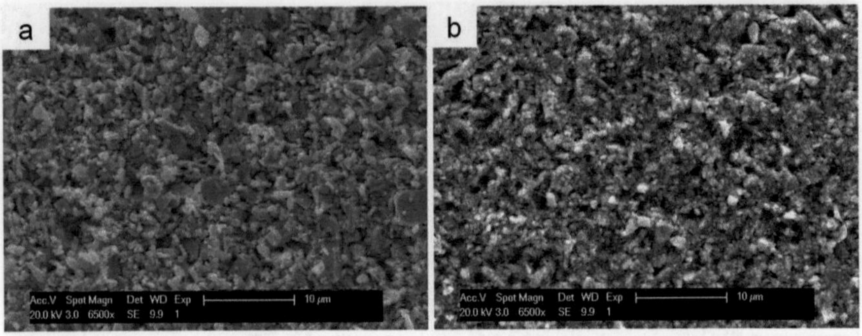

Fig. 3 SEM images of Dy_2O_3 **a** as starting material; **b** after two hours in molten $CaCl_2$ at 1173 K (900 °C)

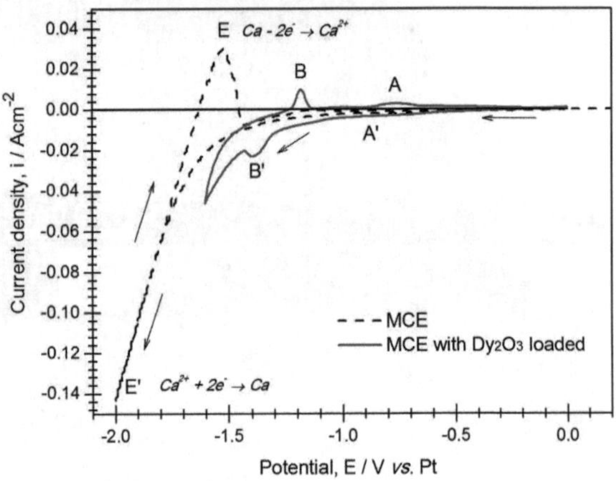

Fig. 4 Cyclic voltammograms in $CaCl_2$ melt on a MCE ($S = 9.3\ 10^{-2}\ m^{-2}$) with and without 0.01 wt% Dy_2O_3, Temperature: 1173 K (900 ° C), scan rate: 50 mVs^{-1}

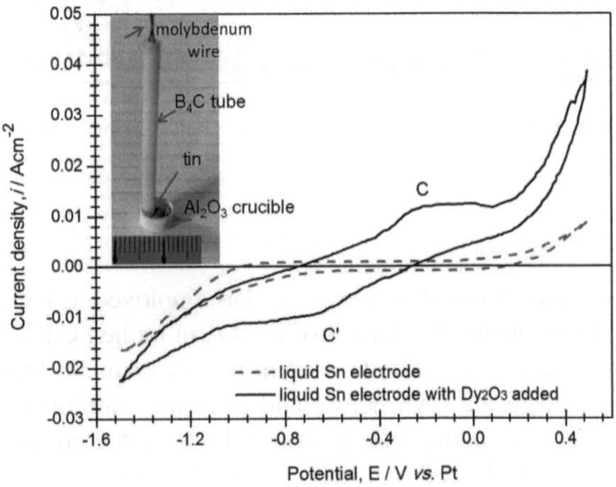

Fig. 5 Cyclic voltammograms of electrode in $CaCl_2$ melt obtained on a liquid Sn electrode ($S = 1.96\ 10^{-3}\ m^{-2}$) with and without 0.01 wt% Dy_2O_3. Temperature: 1173 K (900 °C), scan rate: 50 mVs^{-1}

cathodic direction. The reduction peak A' in the solid curve at about −0.65 V *vs* Pt corresponds to the reduction of Dy(III) ions to Dy at the liquid Sn electrode.

Thus, it can be inferred that the number of electrons exchanged should be three for the Dy(III) reducing on a liquid tin electrode. The electrode reaction can be written as:

$$Dy_2O_3 + 3e^- \leftrightarrow 2Dy + 3O^{2-} \quad (C/C') \tag{3}$$

From what has been discussed above, two completely different mechanisms were observed. There are two reduction steps for Dy_2O_3 on a MCE but only a single step on a liquid Sn electrode. One possible explanation for the difference is that compared to the solid cathode, the metal element can be dissolved in a liquid cathode, which causes a change of the activity and deposition potential of the metal element at liquid cathodes [2].

Conclusions

The dissolution of solid Dy_2O_3 in molten $CaCl_2$ was investigated by CSLM. Results illustrated that it dissolves only slightly within 2 h. The SEM results suggested that Dy_2O_3 is stable in molten $CaCl_2$ at 1173 K (900 °C) and no other compounds (like e.g. DyOCl) were detected after quenching. This make Dy_2O_3 a good candidate for the direct electro-reduction to metal. The electrochemical behavior of dysprosium oxide at a solid molybdenum and a liquid tin cathode were investigated by cyclic voltammetry in a $CaCl_2$ melt at 1173 K (900 °C). Two reduction steps can be noticed for Dy_2O_3 on a solid Mo electrode but only a single step was found on a liquid Sn electrode.

Acknowledgements The authors thank KU Leuven for financial support (IOF-KP Rare[3] project). The authors thank Joris Van Dyck and Joop Van Deursen for their technical assistance.

References

1. Y. L. Yaropolov, A. S. Andreenko, S. A. Nikitin, S. S. Agafonov, V. P. Glazkov, V. N. Verbetsky, "Structure and magnetic properties of RNi (R = Gd, Tb, Dy, Sm) and $R_6M_{1.67}Si_3$ (R = Ce, Gd, Tb; M = Ni, Co) hydrides", *J. Alloys Compd.*, 509(2011), S830–S834.
2. Y. Castrillejo, M. R. Bermejo, A. I. Barrado, R. Pardo, E. Barrado, A. M. Martinez, "Electrochemical behaviour of dysprosium in the eutectic LiCl-KCl at W and Al electrodes", *Electrochim. Acta*, 50(2005),2047–2057.
3. H. Zhu, *Rare earth metal production by molten salt electrolysis, Encyclopedia of Applied Electrochemistry*, (New York, NY, Springer 2014), 1765.
4. G. Z. Chen, D. J. Fray, T. W. Farthing, "Direct electrochemical reduction of titanium dioxide to titanium in molten calcium chloride", *Nature*, 407(2001), 361–364.

5. P. Kim, H. Xie, Y. Zhai, X. Zou, X. Lang, "Direct electrochemical reduction of Dy_2O_3 in $CaCl_2$ melt", *J. Appl. Electrochem.*, 42(4)(2012), 257–262.
6. H. Xie, P. Kim, J. Wang, Y. Zhai, X. Zou, X. Lang, "Study on the electrodeoxidation of solid Dy_2O_3 in molten $CaCl_2$", *Adv. Mater. Res.*, 391(2011), 955–959.
7. S. Feichtinger, S. K. Michelic, Y. Kang, C. Bernhard, "*In-situ* observation of the dissolution of SiO_2 particles in CaO-Al_2O_3-SiO_2 slags and mathematical analysis of its dissolution pattern", *J. Am. Cera. Soc.*, 97(1)(2014), 316–325.
8. C. Orrling, S. Sridhar, A. Cramb, "*In-situ* observation of the role of alumina particles on the crystallization behavior of slags", *ISIJ Int.*, 40(2000), 877–885.
9. F. Verhaeghe, J. Liu, M. Guo, S. Arnout, B. Blanpain, P. Wollants, "Dissolution and diffusion behavior of Al_2O_3 in a CaO-Al_2O_3-SiO_2 liquid: An experimental-numerical approach", *Appl. Phys. Lett.*, 91(2007), 1241041–1241043.
10. F. Verhaeghe, S. Arnout, B. Blanpain, P. Wollants, "Lattice Boltzmann model for diffusion-controlled dissolution of solid structures in multicomponent liquids", *Phys. Rev.*, 72(2005), 0363081–0363084.
11. D. Wang, X. Jin, G. Z. Chen, "Solid state reactions: an electrochemical approach in molten salts", *Annu. Rep. Prog. Chem. Sect. C: Phys. Chem.*, 104(2008), 189–234.
12. M. E. Sibert, Q. H. Mckenna, M. A. Steinberg, E. Wainer, "Electrolytic reduction of titanium monoxide", *J. Electrochem. Soc.*, 102(5)(1955), 252–262.

Part III
Solid-State Processing

Evaluation Feature of Nano Grain Growth of TiO_2 Thin Film via Sol-Gel Route

Habibollah Aminirastabi, Zhangzhao Weng, Zhaoxian Xiong, Goli Ji and Hao Xue

Abstract TiO_2 films were successfully coated via a spin coating sol–gel route, using tetrabutyl titanate as the main raw reagent. One batch of samples was sintered at different temperatures, ranges from 300 to 800 °C, for 2 h, while other ones were treated at 700 °C for different soaking times (1–600 min). The phase compositions of the films were identified with X-ray diffraction (XRD). The microstructure of the coating were observed by a scanning electron microscope (SEM). Experimental results indicated that well-developed crystallinity and good morphology of TiO_2 could be synthesized above 500 °C for 2 h. The crystallinity and morphology of grains in films were affected by the reaction temperature and time. However, image analysis shows that sintering temperature had a great impact on the crystallinity and particles size of the samples above 700 °C. Kinetic exponents of grain growth were also obtained with image processing software.

Keywords TiO_2 · Grain growth · Fractal dimension · Sol-gel route

Introduction

Anatase, brookite and rutile are three polymorphs of TiO_2 found in nature. Brookite is a naturally occurring phase, and is extremely difficult to synthesize, but anatase and rutile can be synthesized in laboratory without difficulty [1, 2].

Titanium oxide (TiO_2) is used for various applications, i.e. photocatalysis, optoelectronics, dye solar cells, gas sensing, luminescent materials, water photolysis, bactericidal use, dielectric materials, cosmetics and pigments, etc. [3–8].

For miniaturization of most of modern systems that use TiO_2, and achieving desirable physical properties, small grains in submicron or nanometer scales are used. In recent years, because of the increased dependence of modern technologies on miniaturization, size effect has found a focal importance in scientific societies

H. Aminirastabi (✉) · Z. Weng · Z. Xiong · G. Ji · H. Xue
College of Materials, Xiamen University, Xiamen 361005, China
e-mail: habib_amini@yahoo.com

A. Allanore et al. (eds.), *Materials Processing Fundamentals 2017*,
The Minerals, Metals & Materials Series, DOI 10.1007/978-3-319-51580-9_4

and research centers. By reducing the size, radical changes appear in the physical properties of ceramics [9, 10].

Depositing of thin films through sol gel processing in wet chemistry is a method that is used in the synthesis of colloidal dispersions of inorganic and organic–inorganic hybrid materials. Sol gel route is capable of producing of nanoparticles, nanorods and thin films. Sol gel process has many advantages, including low processing temperature and homogeneity in molecular levels. In this method, the morphology of microstructure and size of nanocluster can be controlled by hydrolysis and condensation reaction. During sintering processes of sol gel films, final morphology is depended on the kinetics of different fundamental formation processes, e.g., nucleation, coagulation and sintering [11, 12].

Titanium oxide that prepare with sol-gel method usually present an amorphous structure, which must be heated in order to achieve the crystal structure [13–15].

It has been shown that the activity of thin film TiO_2 is influenced by crystal structure, surface area, size distribution, porosity, band gap, and density [16–19]. Retaining nanoscale grain sizes and the rapid grain growth during sintering are the main challenging factors in the sintering process of nanoparticles [20]. Usually, the density increases with increasing sintering time or temperature [21, 22].

The process of grain growth is considerably interconnected with the development and morphology of the pores in grain aggregates. Dimension change of pores during the process of sintering can be an index of grain growth.

Fractal theory is used in many fields, such as physics, chemistry, electro-ceramics, biology and hydrology [23, 24]. There are a lot of papers indicating that the aggregates pores in solids have fractal structures [25–30]. Therefore, the fractal dimension of pores and grain boundaries can be used as an index of grain growth, which can represent a phenomenological description of grain growth in ceramics. Another advantage of the fractal description of grain growth lies on its usage in the aggregations of nano and molecular systems that are too complicated for modeling and understanding their mechanisms.

In this work, we prepared TiO_2 films using sol-gel route, and investigated the influence of sintering conditions, temperature and soaking time, on the grain parameters that can significantly alter the microstructure of the sample. Fractal Dimension, as a microstructure index, describes the relations between grain growth kinetics and grain microstructure, and traces any changes in grain microstructure during sintering process.

Experimental

In this study, TiO_2 films with nanocrystal grains were synthesized by sol-gel route (Fig. 1). Tetrabutyltitanate ($Ti\ (OC_4H_9)_4$) was used as the main raw reagent, and acetic acid (CH_3COOH) and ethyl alcohol as solvents. Acetic acid was

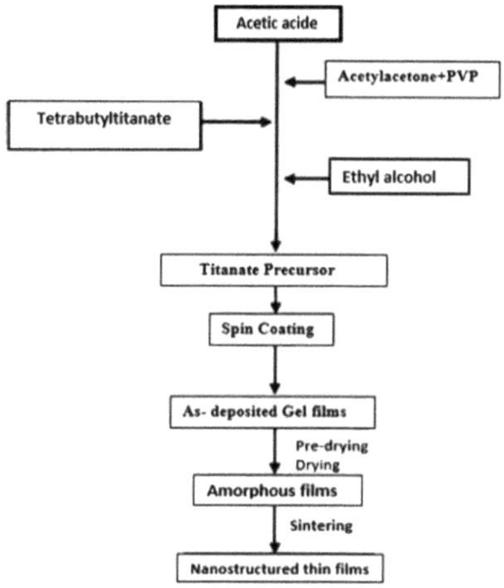

Fig. 1 The preparation process of TiO_2 film by sol gel route

mixed with polyvinylpyrolidone dissolved in ethyl alcohol and acetyl acetate and then Tetrabutyltitanate was added. The mixture was stirred vigorously at room temperature for 1 h to form TiO_2 precursor.

Silicon wafers (coated with ITO (indium tin oxide)) were used as substrates for the deposition of TiO_2 coatings.

Before deposition, the Si wafers were washed with ethyl alcohol, acetone and ammonium to eliminate adsorbed impurities and to form a stable film.

Sol-gel solution was dispersed on the substrates and then spin coated at 3000 rev./min for 60 s. One batch of samples were sintered at 300, 400, 500, 600, 700 and 800 °C for 2 h to study microstructure at different temperature. To study the changes in microstructure during the isothermal hold (700 °C), other samples were held for 1, 3, 6, 10, 30, 60, 100, 300, 600 min's, and then allowed the furnace to cool. To study the sintering process in stages, samples were heated in atmospheric air with a heating rate of 3 °C/min in conventional furnace.

The crystallization process of TiO_2 gel was studied with the aid of DSC/TG (German NETZSCH, STA449 F3), and XRD (powder diffractometer). The microstructure of the powders was examined with SEM (Hitachi SU-70). The grain size of TiO_2 nanocrystals was calculated with image analysis IMAGE J and fractal dimension was calculated with FRACTAL ANALYSIS SYSTEM version 3.4.7.

Results and Discussion

Characterization of TiO_2 Nanocrystals

The thermal decomposition process of the TiO_2 Sol-gel after preheating (40 °C, 24 h) and during calcination in air was studied using DTA/TG. The results are shown in Fig. 2. As can be seen, the total weight loss of the sol-gel from 30 to

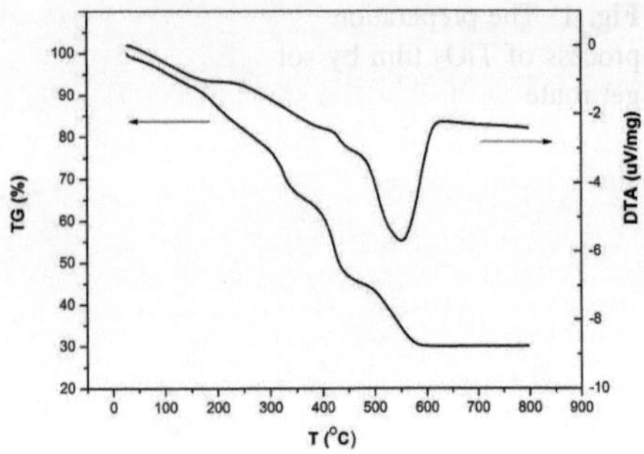

Fig. 2 DTA/TG curves of the TiO_2 nanocrystal precursor of the sample

800 °C amounts to 70%. The decomposition process can be divided into four stages. The temperature range in the first, second and third stage is 30–110 °C, 100–250 °C and 250–450 °C, respectively. The weight loss in each stage is about 5.2, 13.77, and 35.05%, which can be correspondingly attributed to the removal of water, the evaporation of organic solvents and the combustion of the organic materials of the framework of the sol-gel. On the DTA curve, no strong significant thermal effect was observed in the second and third stage. The last stage is in the temperature range of 450-800 °C. The corresponding weight loss is 16.66%. On the DTA curve, an endothermic peak at 550 °C appears, which signifies the formation

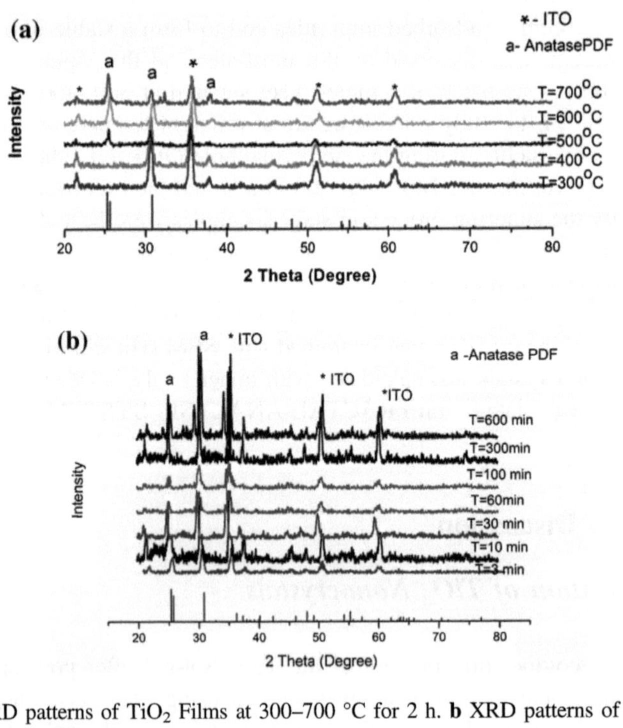

Fig. 3 **a** XRD patterns of TiO_2 Films at 300–700 °C for 2 h. **b** XRD patterns of TiO_2 films at 700 °C from 3 to 600 min

of TiO_2 or phase transition. The weight loss in this stage is probably due to the removal of carbon materials.

XRD patterns of the TiO_2 film obtained via sintered at different conditions are shown in Fig. 3a and b. With increasing calcination temperature from 300 to 700 °C (duration = 2 h), the XRD results of TiO_2 film shows the peak ratio of XRD patterns between anatase and ITO increased.

Figure 3a indicates that the calcination temperature (300–500 °C with 2 h soaking time) is not high enough that the TiO_2 grain be completely sintered.

By increasing soaking time from 3 min to 600 min (T = 700 °C), Fig. 3b indicates that upon calcination at 700 °C for more than 100 min's, the peak intensity is sharper and narrower than that of the times less than 100 min. This implies that thermal aging (>100 min) at a temperature of as high as 700 °C is necessary for the completely crystallization of TiO_2 and the complete removal of carbon from the product. (For comparison, the peak ratio of XRD patterns between anatase and ITO (indium tin oxide) are used).

Grain Growth Kinetics of TiO_2 Nanocrystals

Grain Growth During Heat-up and 2 H Soaking Time

The morphology of the nano-size titanate is shown in Fig. 4a. Images of layers in 300–700 °C are obtained in bulk of layers and not exactly clear because structure of layers between these temperatures have large amount of polymer.

Using the image analysis technique, the average grains size of TiO_2 film calculated. After sintering at 300–800 °C with soaking time 2 h, average grains sizes were 25–190 nm, approximately.

The plots of fractal dimension and grain size during heat-up with 2 h soaking time is displayed in Fig. 4b and c.

The rapid grain growth during this stage of sintering is easily observed from Fig. 4a and b after sintering at 800 °C.

Figure 4c shows variation of fractal dimension of nano structure TiO_2 at different temperatures with 2 h soaking time. According to this figure, surface of layer will be smooth after heat treatment at 600 °C, and hence leads to fractal dimension decreased. It can be seen, the size of fractal dimension is not depended on area selected (every box in Fig. 4c shows pixel of the selection area).

Fractal Dimension of Grain Boundary

During the grain growth process, the grain area and grain perimeter in two-dimension can be obtained and with fractal dimension, the volume of grain and the surface of grain boundary in three dimension can also be obtained [19, 20].

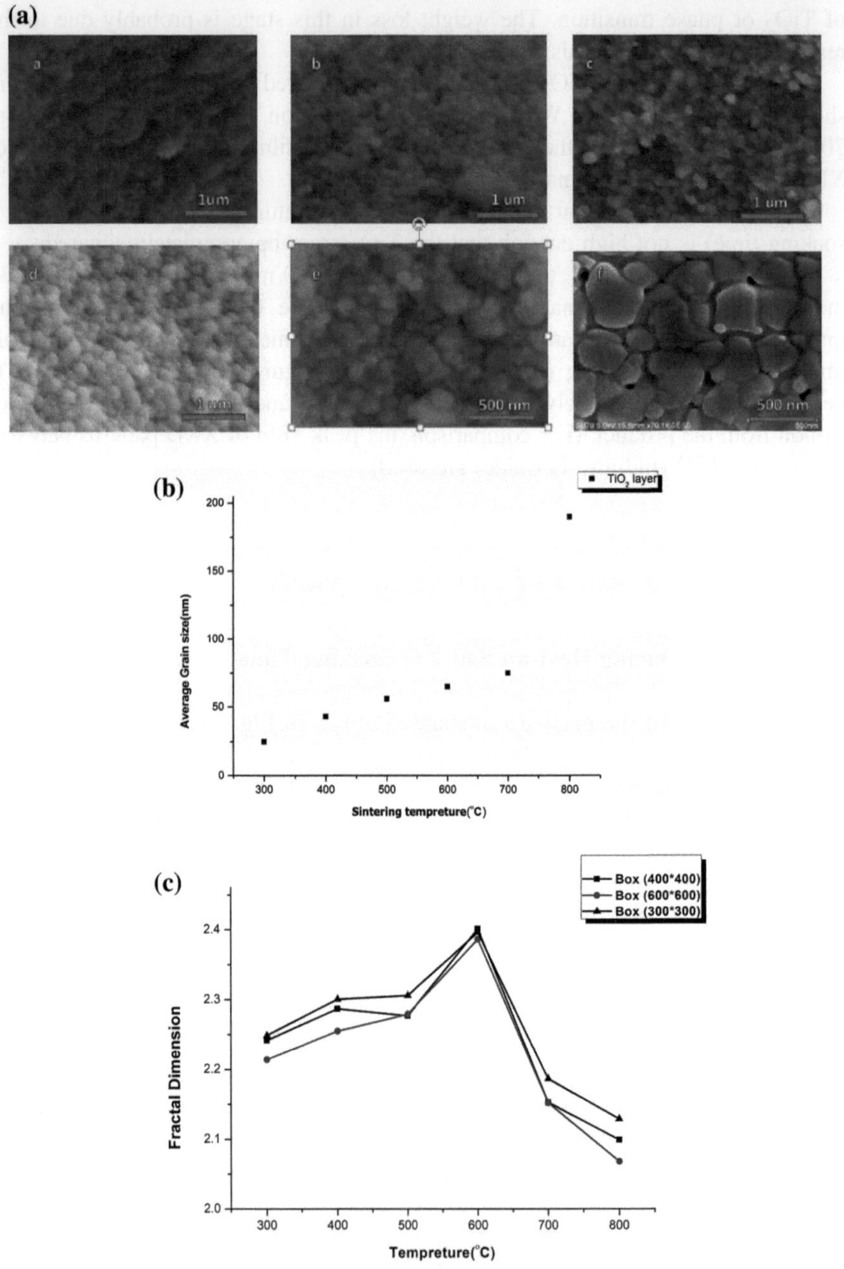

Fig. 4 **a** SEM micrographs of TiO_2 sintered at (*a*) 300 °C, (*b*) 400 °C, (*c*) 500 °C, (*d*) 600 °C, (*e*) 700 °C, and (*f*) 800 °C for 2 h. **b** TiO_2 grain growth during heat up at 300–800 °C for 2 h soaking time. **c** Variation of fractal dimension of nano structure TiO_2 at different temperatures for 2 h

$$P_G^{1/D} \propto A_G^{1/2} \quad (1)$$

Here, P_G is the perimeter of grain boundary and A_G is the area of grain;

$$A_G^{1/D} \propto V_G^{1/3} \quad (2)$$

where A_G is the surface area of grain boundary and V_G is the volume of grain. They can be transformed and simplified into logarithmic forms,

$$\mathrm{Log}\, P_G = D/2 \log A_G + C_1 \quad (3)$$

And

$$\mathrm{Log}\, A_G = D/3 \log V_G + C_2. \quad (4)$$

Fractal dimension (D) is calculated by the method of box-counting after pre-processing SEM images.

Grain Growth During Isothermal Holdings

In order to understand the entire grain growth process, isothermal sintering experiments were carried out at 700 °C, following heat-up at 3 °C/min. Figure 5a shows the isothermal grain growth curve of grain size versus time at 700 °C for all samples with same initial grain sizes. It demonstrates that the grain growth behavior of isothermal hold follows a more gradual process than that of heat-up.

Kinetic Exponent of Grain Growth

Generally, grain growth occurs in polycrystalline materials to decrease the free energy of the system by decreasing the total grain boundary energy. The grain growth behavior of polycrystalline materials was reported to obey the equation: The rate of grain growth, is inversely proportional to grain size:

$$d^n - d_o^n = (c)t \quad (5)$$

where, d is the average grain size at time t, d_o is the initial average grain size, n the grain growth exponential reflecting grain growth behavior, and c is the temperature-dependent rate constant can be expressed by the Arrhenius equation

$$c = Se^{(-E/RT)} \quad (6)$$

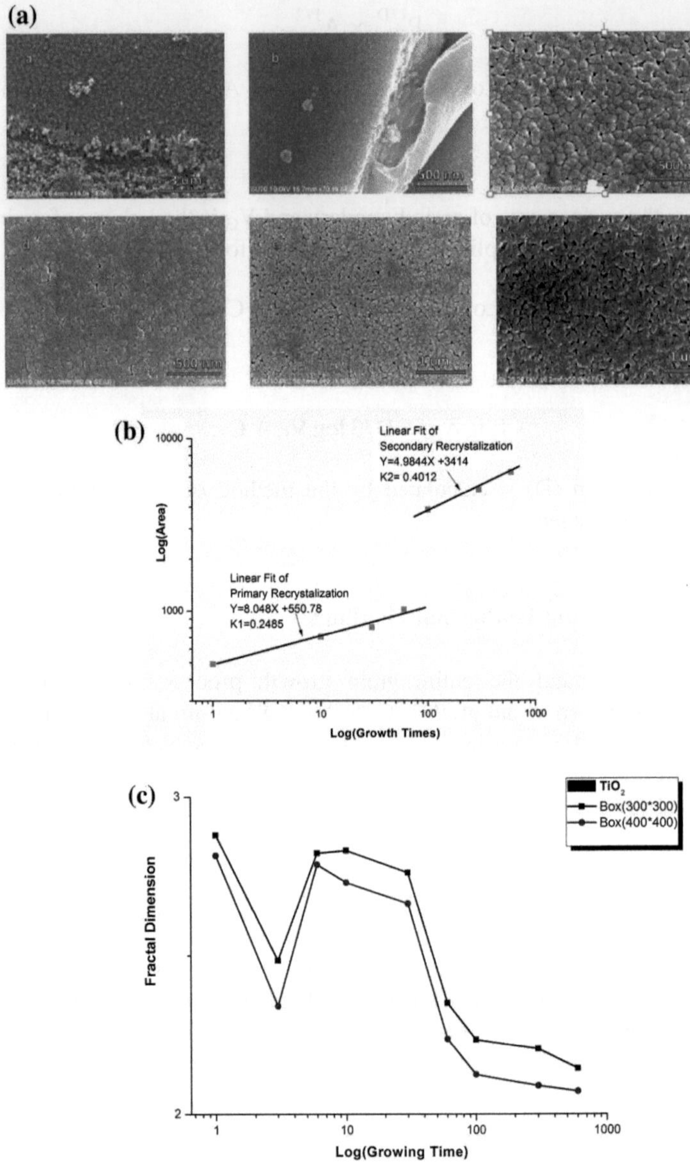

Fig. 5 **a** SEM micrograph of TiO_2 film sintered at 700 °C for (*a*) 1 min, (*b*) 3 min, (*c*) 60 min, (*d*) 100 min, (*e*) 300 min, (*f*) 600 min. **b** Grain area versus growth times TiO_2 film at 700 °C. **c** Variation of fractal Dimension of nano structure TiO_2 sintered at different socking time

s is the pre-exponential factor, *E* is the activation energy, *R* is the universal gas constant. Since d_o is generally much smaller than d Eq. (5), Eq. (6) can be derived as follows:

$$\text{Log A} = (2/k)\,\text{Log(t)} + C_3 \tag{7}$$

k is the kinetic exponent of grain Growth, A is the average grain area, and C_3 is a constant. K can be obtained from the slope of the linear regressive line of log A versus log t.

As all of grain areas can be collected during the sintering process, average grain areas in different grains can be calculated. Therefore, the average grain area versus growth time can be plotted in a logarithmic scale. The kinetic exponents of grain growth correspond inversely to the slope of curve. They are about 0.24 at the early stages of the sintering process and then increase to 0.40 at the later stages, as shown in Fig. 5b. It can be seen that the kinetic exponents of grain growth are actually a function of the time. At the early stage of sintering has smaller activation energy than those at the later stage of process.

According to Fig. 5c, the variation of fractal dimension of nano structure TiO_2 sintered at different socking times, the fractal dimension decrease with the time of grain growth. This means that the surfaces of grain boundaries become smoother with growth time.

Conclusion

Titanium oxide films were obtained successfully by sol gel route. The results of TG and DTA indicate that decomposition of sol gel proceeds in four stages. XRD results of titanium oxide film show the main diffraction peaks at 700 °C (duration = 2 h) become narrower and sharper than that prepared at lower temperature. The grain growth behavior of isothermal hold follows a more gradual process than that of heat-up.

The kinetic exponent of grain growth in the secondary recrystallization is larger than that of primarily recrystallization at 700 °C.

The fractal dimension of microstructure ceramics will be changed during different temperature and soaking times, and therefore, the fractal dimension can be taken as an index of grins growth.

Acknowledgements Authors appreciate Prof Suzuki for providing the Software of FRACTAL ANALYSIS SYSTEM, version 3.4.7, and, thanks to supporting financial by Fujian Provincial Science and Technology Department with project number of 2013H6020.

References

1. P Y Simons, F Dachille, “The structure of TiO_2 II, a high-pressure phase of TiO_2”, Acta Crystallographic, 2001, 23(2):334–336.
2. L J Alemany, M A Bañares, E Pardo, et al. “Morphological and Structural Characterization of a Titanium Dioxide System”, Materials Characterization, 2000, 44(3):271–275.

3. J Chen, H Wang, X Wei, et al. "Characterization, properties and catalytic application of TiO_2, nanotubes prepared by ultrasonic-assisted sol-hydrothermal method", Materials Research Bulletin, 2012, 47(11):3747–3752.
4. G Liu, K Wang, N Hoivik, "Small diameter TiO_2, nanotubes with enhanced photo responsivity", Electrochemistry Communications, 2013, 28(1):107–110.
5. C H Han, H S Lee, S D Han, "Synthesis of Nanocrystalline TiO_2 by Sol-Gel Combustion Hybrid Method and Its Application to Dye Solar Cells", Bulletin- Korean Chemical Society, 2008, 29(8):1495–1498.
6. Z Wang, L Shi, F Wu, et al. "The sol-gel template synthesis of porous TiO_2 for a high performance humidity sensor", Nanotechnology, 2011, 22(27).
7. M. Radetić, "Functionalization of textile materials with silver nanoparticles", Journal of Photochemistry & Photobiology C Photochemistry Reviews, 2013, 16(1):62–76.
8. K Ohenoja, M Illikainen, J Niinimäki, "Effect of operational parameters and stress energies on the particle size distribution of TiO2 pigment in stirred media milling", Powder Technology, 2013, 234(1):91–96.
9. J Zhu, C Jin, W Cao, et al. "Phase transition and dielectric properties of nanograin $BaTiO_3$ ceramic under high pressure", Applied Physics Letters, 2008, 92(24).
10. H Ghayour, M Abdellahi, "A brief review of the effect of grain size variation on the electrical properties of $BaTiO_3$-based ceramics", Powder Technology, 2016, 292:84–93.
11. C J Brinker, G W Scherer, "Sol-Gel Science", Academic Press, 1990:475–495.
12. J D Wright, N A J M Sommerdijk, "Sol-Gel Materials: Chemistry and Applications", Gordon and Breach Science Publishers, 2001.
13. C Hu, S Duo, R Zhang, et al. "Nanocrystalline anatase TiO_2, prepared via a facile low temperature route", Materials Letters, 2010, 64(19):2040–2042.
14. P P Lottici, D Bersani, M Braghini, et al. "Raman scattering characterization of gel-derived titania glass", Journal of Materials Science, 1993, 28(28):177–183.
15. P Hartmann, D K Lee, B M Smarsly, et al. "Mesoporous TiO_2: Comparison of Classical Sol − Gel and Nanoparticle Based Photoelectrodes for the Water Splitting Reaction", Acs Nano, 2010, 4(6):3147–54.
16. H. V. Damme, "Photocatalysis - fundamentals and applications", In Photocatalysis - Fundamentals and Applications, N. Serpin and E. Pelizzetti, eds., Wiley, New York, Chapt. 7 (1981).
17. H Harada, T Ueda, "Photocatalytic activity of ultra-fine rutile in methanol-water solution and dependence of activity on particle size", Chemical Physics Letters, 1984, 106(3):229–231.
18. R. I. Bickley, T. González-Carreño, J. S. Less, L. Palmisano, and R. J. D. Tilley, "A structural investigation of titanium-dioxide photocatalysts", Solid State Chem. 1991: 92:178.
19. K Kobayakawa, Y Nakazawa, M Ikeda, et al. "Influence of the Density of Surface Hydroxyl Groups on TiO_2, Photocatalytic Activities", Berichte Der Bunsengesellschaft Für Physikalische Chemie, 1990, 94(12):1439–1443.
20. J. E. Kim, C. H. Song, and H. W. Park, "Kinetics of crystallization in $4BaTiO_3$-SiO_2 glass", Mater. Sci. Eng. Volumes 449–451, 25 March 2007, Pages 299–301.
21. J Das, K Chandra, P S Misra, et al. "Novel powder metallurgy technique for development of Fe–P-based soft magnetic materials", Journal of Magnetism & Magnetic Materials, 2008, 320 (6):906–915.
22. G V Samsonov, V I Yakovlev, "Activated sintering of tungsten with nickel additions", Soviet Powder Metallurgy & Metal Ceramics, 1967, 6(8):606–611.
23. K N Ramakrishnan, S Venkadesan, K P N Murthy, "A fractal description of grain boundaries in a sintered powder metallurgical sample", Scripta Metallurgica et Materialia, 1995, 32 (32:5):685–688.
24. G Reisel, R B Heimann, "Correlation between surface roughness of plasma-sprayed chromium oxide coatings and powder grain size distribution: A fractal approach", Surface & Coatings Technology, 2004, 185(2–3):215-221.
25. A J Hickey, N M Concessio, "Descriptors of irregular particle morphology and powder properties", Advanced Drug Delivery Reviews, 1997, 26(1):29–40.

26. T Akiyama, T Iguchi, K Aoki, et al. "A fractal analysis of solids mixing in two-dimensional vibrating particle beds", Powder Technology, 1998, 97(1):63–71.
27. C Y Poon, R S Sayles, T A Jones, "Surface measurement and fractal characterization of naturally fractured rocks", Journal of Physics D Applied Physics, 1992, 25(8):1269–1275(7).
28. L C Y Chan, N W Page, "Particle fractal and load effects on internal friction in powders", Powder Technology, 1997, 90(3):259–266.
29. N Bird, M C Díaz, A Saa, et al. "Fractal and multifractal analysis of pore-scale images of soil", Journal of Hydrology, 2006, 322(1–4):211–219.
30. A Dathe, M Thullner, "The relationship between fractal properties of solid matrix and pore space in porous media", Geoderma, 2005, 129(3–4):279–290.

Part IV
Molten & Gas State Processing

Influence of Oxygen on Surface Tension of Zirconium

Jie Zhao, Jonghyun Lee, Rainer Wunderlich, Hans Fecht, Stephan Schneider, Michael SanSoucie, Jan Rogers and Robert Hyers

Abstract Zirconium samples with different oxygen concentrations were tested using a ground-based electrostatic levitator at NASA Marshall Space Flight Center. The surface tension of liquid zirconium samples was measured in both undercooled and superheated conditions. The effect of oxygen on surface tension was determined: oxygen in zirconium samples decreases the surface tension of liquid samples, with only a small change in the temperature dependence.

Keywords Electrostatic levitation · Surface tension · Oxygen · Zirconium

Introduction

The influence of oxygen on the thermophysical properties of zirconium is being investigated using MSL-EML (Material Science Laboratory Electromagnetic Levitator) on ISS (International Space Station) in collaboration with NASA, ESA (European Space Agency), and DLR (German Aerospace Center). Zirconium samples with different oxygen concentrations are being processed through multiple melt cycles, during which the density, viscosity, surface tension, heat capacity, and

J. Zhao · J. Lee (✉) · R. Hyers
Department of Mechanical and Industrial Engineering, University of Massachusetts, Amherst, MA, USA
e-mail: jonghyunlee@ecs.umass.edu

R. Wunderlich · H. Fecht
Institute of Micro and Nanomaterials, Ulm University, Ulm, Germany

S. Schneider
Institute for Materials Physics, German Aerospace Center (DLR), Cologne, Germany

M. SanSoucie · J. Rogers
NASA Marshall Space Flight Center, Huntsville, AL, USA

A. Allanore et al. (eds.), *Materials Processing Fundamentals 2017*,
The Minerals, Metals & Materials Series, DOI 10.1007/978-3-319-51580-9_5

electric conductivity are being measured at various undercooled and superheated temperatures. The facility check-up of MSL-EML and the first set of melting experiments were successfully performed in 2015. The first zirconium sample was tested at the end of 2015. As part of ground support activities, the thermophysical properties of zirconium and Zr-O were measured using a ground-based electrostatic levitator located at the NASA Marshall Space Flight Center. The influence of oxygen on the measured surface tension was evaluated over a wide range of temperature. The results of this research will serve as reference data for comparison to those measured in ISS.

Electrostatic levitation (ESL) is a containerless method for processing materials including metallic glasses, quasicrystal-forming alloys, and industrial alloys such as steels, semiconductors and refractory metals. Compared with samples used by EML, samples in ESL experiments are smaller with typically 2–3 mm diameter (6–8 mm in EML). The sample is electrically charged by a D_2 arc lamp and levitated in an electrostatic field generated by upper and lower electrodes. Figure 1 show a levitated sample during measurement and Fig. 2 gives a schematic ESL system [1]. The sample's position is maintained by two dual-axis position detectors and a

Fig. 1 Sample levitated between electrodes and heated by laser [1]

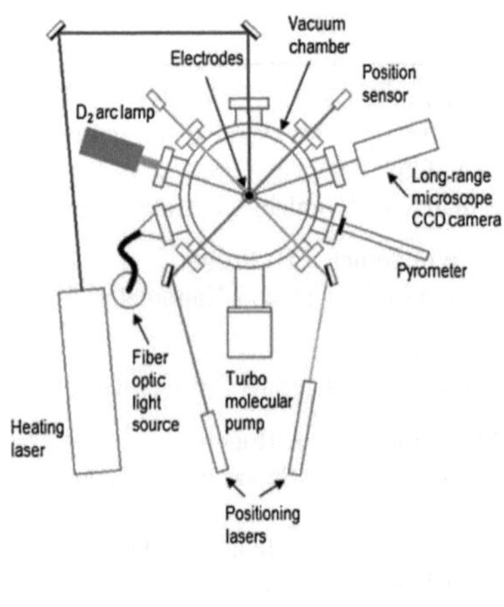

Fig. 2 Schematic of ESL system [1]

feedback system. The charge on samples can be controlled by an UV source via the photoelectric effect. Samples can be levitated as long as charged. To achieve desired temperatures in study the sample is heated by a 200 W Nd:YAG laser. A radial camera records the processing video of sample. The electrostatic levitator facilities require a non-contact method for measuring temperature. ESL experiments use optical pyrometers to measure thermal radiation emitted by sample. Then the temperature can be determined by the emissivity of the sample. The emissivity of a few materials can be measured directly at a common wavelength in NASA MSFC (Marshall Space Flight Center) ESL. [1] However, the emissivity value of the tested Zr samples was calibrated using the phase transition temperatures, α-Zr to β-Zr at 1136 K.

Surface Tension and Viscosity Measurements

It is difficult to measure the surface tension and viscosity of molten metals or alloys by traditional measurement techniques due to the high temperature. Another difficulty is that the sample in liquid phase tends to be contaminated easily. Electrostatic levitation (ESL), has been used by several groups [2, 3] to measure the thermophysical properties of molten alloys, which eliminates contamination by a container. Since the sample is positioned and heated by two independent systems, and the electrostatic system does not generate any heat, a wide temperature range can be achieved. The levitated sample can be undercooled or superheated and sustain the state long enough to complete the measurement.

Measurements of surface tension and viscosity share the same oscillating drop method of containerless processing. In this work, the oscillating droplet technique was adopted to measure the surface tension and viscosity of liquid zirconium. The levitated sample can be melted and oscillated in a force-free environment as a high-quality mechanical oscillator. The surface tension is related to the resonant frequency of the oscillation. The resonant frequency of the liquid metal sphere is related to the surface tension of the liquid metal sphere, which can be evaluated by Rayleigh's equation [4]:

$$f^2 = \frac{l(l-1)(l+2)\gamma}{3\pi M}$$

where f is the cycle frequency of oscillation; γ is the surface tension of the metal droplet; l is the oscillation mode number which is 2 in this case. M is the droplet mass, which was scaled before and after processing and show small mass loss (Table 1). Thus, the sample evaporation during the process was neglected.

In this method, the levitated liquid sample is excited to oscillate with an additional alternating electrostatic field near the natural frequency of the sample. The oscillating sample finds its natural frequency as it dampens freely to its equilibrium state. The surface tension is determined by the oscillation frequency of the sample.

Table 1 Sample mass loss in ESL

Composition	Samples	Pre-test (mg)	Post-test (mg)	Mass loss (%)
57 ppm	#062	44.514	44.459	0.124
	#069	44.511	44.492	0.043
	#067	43.288	43.273	0.035
1560 ppm	#073	43.266	43.256	0.023
	#071	48.769	48.753	0.033
	#044	47.263	47.226	0.078
2.42%	#048	44.427	44.398	0.065
	#050	46.964	46.930	0.072
	#058	46.101	46.069	0.069
Average				**0.060**
1 Standard deviation				0.029

The measurement of surface tension can be affected by the Marangoni flow on the surface of the liquid sample. For example, in ESL experiments, when the period of recirculation of the flow loops is comparable to that of oscillation, the sample can resonate and self-oscillate. This phenomenon increases errors for surface tension measurements.

Experiments

In the present work, the surface tension of zirconium samples with different oxygen concentrations was measured by the oscillation droplet method using ESL at NASA MSFC. After being levitated and melted, the sample was excited by electrostatic field and damped freely to its equilibrium state. The oscillating frequency was extracted from video data to determine the surface tension.

The three oxygen concentrations of zirconium samples include zirconium with 57 ppm oxygen, and zirconium with 2.42 at.% oxygen. To prepare samples, raw materials were bought from Alfa Aesar. After a hole was drilled on one side of a 6.35 mm diameter × 12.7 mm length zirconium slug, zirconia powder was loaded in the hole and the hole was capped with a 3.175 mm diameter × 3.175 mm length zirconium slug. Then the capped slug was arc-melted into an ingot. The ingot was cut into two halves. One of the halves was sent to Luvak Inc. Boylston, MA for elemental analysis (Method: Inert gas fusion—ASTM E 1019-11). The other was potted in epoxy and cut into several 30–60 mg pieces using a diamond saw. The cut pieces were arc melted again to form near-spherical shape. The preparation process is shown in Fig. 3.

The zirconium samples with different oxygen concentrations were put into multiple melt cycles. A solid sample was heated and melted by a heating laser in vacuum chamber with pressure 10E-7 Torr. As soon as the sample temperature

Fig. 3 Sample preparation process

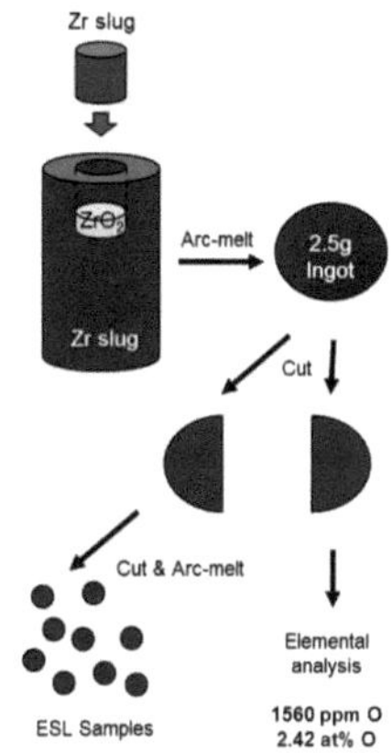

reaches the desired superheated temperature, about 100 K above melting point, the heating laser was modulated to put the sample on a thermal hold. The sequence of the thermal hold was randomized to minimize the effect of mass evaporation on the measured surface tension. A typical temperature-time plot is shown in Fig. 4. The mass loss of the tested samples is shown in Table 1 and it turns out to be less than 0.124% and average is 0.06%.

Figure 5 shows the change in the projected area in pixels of a sample during excitation and damping. Between the two vertical lines is the damping region, which is mixed with some noise. Since the Rayleigh equation [4] was derived based on assumptions that there is no external force and the amplitude is very small, more

Fig. 4 Temperature versus Time in one melt cycle. The *blue straight lines* give the number of oscillations the sample completed

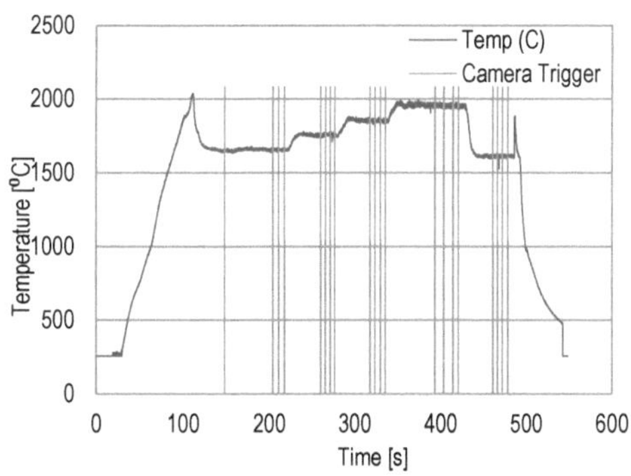

Fig. 5 Sample oscillation and free damping

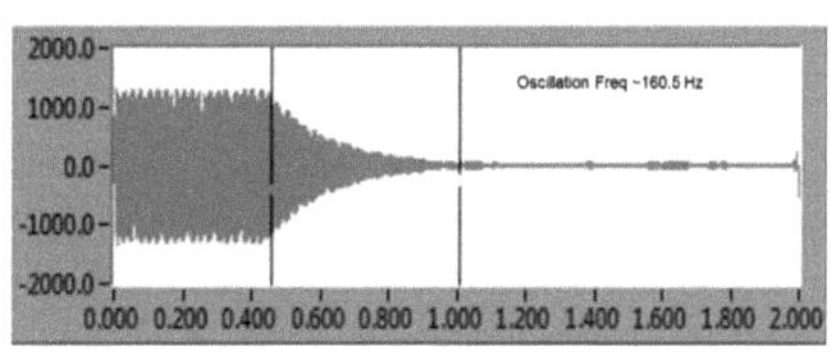

accurate surface tension can be determined if the latter portion of the damping region is to be used.

Experimental Results

The measured surface tension values are plotted in Fig. 6. There are three different oxygen concentrations of zirconium samples—57 ppm, 1560 ppm and 2.42 at.%, with three samples for each concentration. The data from the same composition shows consistent results. The difference between Zr + 57 ppm and Zr + 1560 ppm oxygen was within the standard error of the measurement, 0.87%, and was determined by a t-test to be not statistically significant. The surface tension of samples with 2.42% oxygen was decreased by 9%. A linear interpolation between the 57 ppm and high oxygen results would give a change in surface tension of 0.58% at 1560 ppm, consistent with the observation that the change is less than 0.87%.

The Fig. 7 shows the present surface tension data of zirconium sample with 57 ppm oxygen was compared with those from Iida and Guthrie [5], Paradis et al. [3] and Rhim et al. [6]. All data show a good agreement one another. The relative errors are less than 0.87%. The surface tension is dependent on oxygen content, changing 3.72% per %O. The Fig. 8 gives the comparison with Iida and Guthrie [5] and Roesner-Kuhn's [7] data with the error bar of the present surface tension. A fair agreement can be observed between the present work and Iida. However, so called a 'boomerang effect' (positive slope at low temperatures and negative at high) which had been reported by Roesner-Kuhn was not observed in the present data.

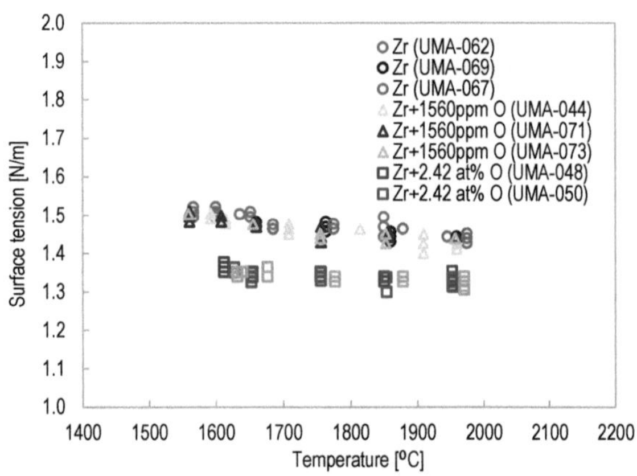

Fig. 6 Surface tension of Zr + 57 ppm, Zr + 1560 ppm oxygen and Zr + 2.42% oxygen + 57 ppm

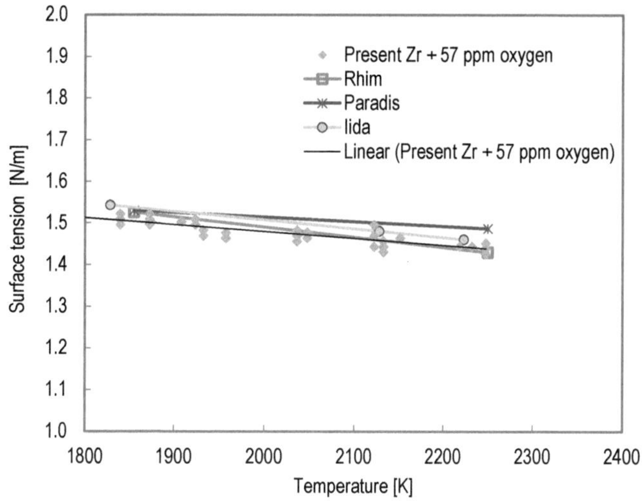

Fig. 7 Surface tension of Zr + 57 ppm oxygen compared with and Iida and Guthrie [5], Paradis et al. [3] and Rhim et al. [6]

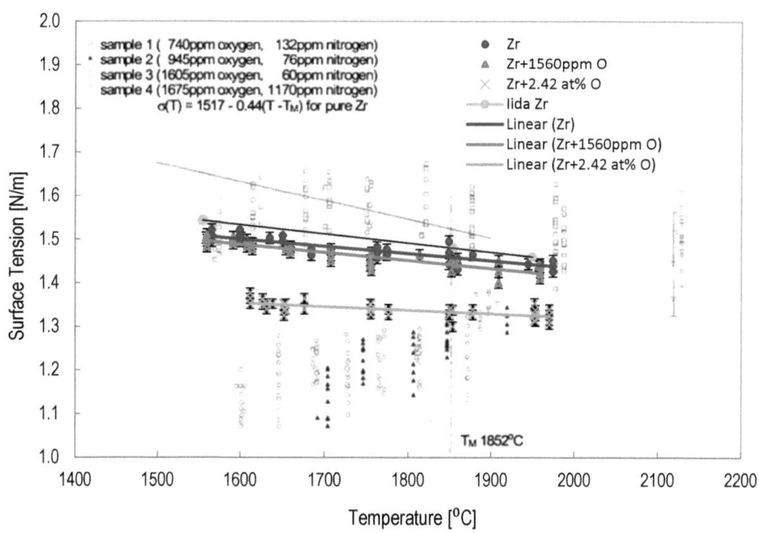

Fig. 8 Surface tension of Zr + 57 ppm oxygen, Zr + 1560 ppm oxygen and Zr + 2.42% oxygen compared with Roesner-Kuhn et al. [7]

Conclusion

The surface tension of Zr-O alloys was measured by a non-contact method. There was excellent agreement with literature data for Zr + 57 ppm oxygen, confirming the accuracy of our measurements. The samples of 57 ppm and 1560 ppm O show no statistically significant change in surface tension within the resolution of our experiment, 0.87%. This result is consistent with the measurement at higher oxygen content. For higher oxygen, the dependence was 3.72%/%O.

References

1. Hyers, R.W. and J.R. Rogers, *A review of electrostatic levitation for materials research.* High Temperature Materials and Processes, 2008. **27**(6): pp. 461–474.
2. Hyers, R.W., *Fluid flow effects in levitated droplets.* Measurement science and Technology, 2005. **16**(2): p. 394.
3. Paradis, P.-F., T. Ishikawa, and S. Yoda, *Non-contact measurements of surface tension and viscosity of niobium, zirconium, and titanium using an electrostatic levitation furnace.* International journal of thermophysics, 2002. **23**(3): pp. 825–842.
4. Rayleigh, L. *On the capillary phenomena of jets.* in *Proc. R. Soc. London.* 1879.
5. Iida, T. and R.I. Guthrie, *The physical properties of liquid metals.* Clarendon Press, Oxford, UK, 1988.
6. Rhim, W.-K., et al., *Noncontact technique for measuring surface tension and viscosity of molten materials using high temperature electrostatic levitation.* Review of scientific instruments, 1999. **70**(6): pp. 2796–2801.
7. Rösner-Kuhn, M., et al., *Investigations of the influence of oxygen on the surface tension of zirconium by the oscillating drop technique.* Surface science, 1999. **443**(3): pp. 159–164.

Thermal Analyses of Silver-Based Sulfosalts in Air

Fiseha Tesfaye and Daniel Lindberg

Abstract The sulfosalts $Ag_{0.93}Cu_{1.07}S$ (stromeyerite) and α-$AgBiS_2$ (schapbachite) have been studied under oxidizing conditions at elevated temperatures. The sulfosalts were synthesized from the pure simple sulfides in evacuated quartz ampoules. The synthesized samples were thermally analyzed in the temperature range from 298 to 1173 K by a simultaneous DTA-TGA analyzer. Based on the DTA measurements the phase transition of stromeyerite to the solid solution $(Cu, Ag)_2S$(hcp) in air is determined to be $T = (360.9 \pm 2)$ K. For the first time, maximum thermal stability of $(Ag, Cu)_2S$ in an oxidizing atmosphere close to the partial pressure of oxygen in air ($P(O_2) \approx 0.20$ atm) is determined to be below $T = (614 \pm 2)$ K, above which it oxidizes to form Ag, CuO and Ag_2SO_4. The melting temperature of Ag_2SO_4 determined from the cooling DTA curve, $T = (932.76 \pm 2)$ K, is in good agreement with the literature value. Below $T = 1173$ K, the oxidation process for schapbachite in air has been indirectly determined to be: $2AgBiS_2 + 5.5O_2(g) \rightleftarrows 2Ag + Bi_2O_3 + 4SO_2(g)$.

Keywords Silver · Sulfosalts · Thermal stability · Oxidation · DTA-TGA analysis · Pyrometallurgical roasting

Introduction

Currently, most of the available sulfide ores are low-grades and also becoming increasingly complex. Ag and Cu bearing sulfosalts are frequently encountered in the available complex ore minerals for the production of base metals and are also common sources of Ag. For instance, in the precious and base metal ores, $AgBiS_2$ is a relatively common sulfobismuthide mineral. Several samples of hydrothermal vein-type ores including matildite (β-$AgBiS_2$) and schapbachite (α-$AgBiS_2$) are

F. Tesfaye (✉) · D. Lindberg
Johan Gadolin Process Chemistry Centre, Åbo Akademi University, Biskopsgatan 8, FI-20500 Turku, Finland
e-mail: fiseha.tesfaye@abo.fi

A. Allanore et al. (eds.), *Materials Processing Fundamentals 2017*,
The Minerals, Metals & Materials Series, DOI 10.1007/978-3-319-51580-9_6

reported to occur along the Northern Kinzigtal Fault (NKF) in the Schwarzwald ore district and in south western Germany [1]. Mineralogical studies of the Cu-Ag deposit in central Europe have also shown that schapbachite (α-$AgBiS_2$) exists commonly in the mineral assemblages for the production of valuable metals [2]. Among the Ag-Cu-based sulfosalts, stromeyerite ($Ag_{0.93}Cu_{1.07}S$) is relatively common ore mineral in most hydrothermal vein and replacement deposits [3]. In order to modify operating flow sheets and strategies for processing more complex feed materials economically, valuable metal producers are in need of accurate thermal stability data of the phases, which are common in these complex ore minerals, at the pyrometallurgical processing conditions.

In recent years, the ternary chalcogenide semiconducting compound $AgBiS_2$ has been studied intensively due to its unique electronic and magnetic properties which can be applied in linear, nonlinear, optoelectronic, and thermoelectric devices as well as optical recording media [4–12]. Recently, the high ionic conductivity and gradual disorder in the sequential phase transitions that are observed in the Ag-Cu-S ternary phases renewed the interest in the silver and copper based sulfides [13]. For instance, the ionic conductivity of superionic solid solution of Ag-Cu-sulfides at $T = 573$ K and composition of Ag/Cu = 1, i.e. $(Cu, Ag)_2S$(fcc) phase, has been reported to be as high as 2.38 S cm^{-1} [14]. Furthermore, the solid solution $(Cu, Ag)_2S$ may have also unique optical properties [15]. There is also growing interest in the development of novel inorganic materials for use in transistors and semiconductor switches, which can be operated with a temperature gradient in air. In order to optimize manufacturing of devices and accurately determine the user limits for devices incorporating the ternary phases, thermal stabilities of the phases at elevated temperatures in air is important. Consequently, data about thermal stabilities of the Ag-based sulfosalts at elevated temperatures in an oxidizing atmosphere is not only crucial for studies of ore genesis, processing of complex minerals and the optimization of the extractive metallurgy of the valuable metals, but also for improved manufacturing and user limit determination of novel electronic materials incorporating the Ag-(Bi, Cu)-S ternary phases. Therefore, the purpose of this experimental investigation was to study the thermal stabilities of the $AgBiS_2$ and AgCuS at elevated temperatures in an oxidizing atmosphere ($P(O_2) \approx 0.20$ atm).

Phase Stabilities of $AgBiS_2$ and AgCuS

Below its congruent melting point at $T_m = (1074 \pm 4)$ K, $AgBiS_2$ exists in two structural forms [16]. The low-temperature phase β-$AgBiS_2$, which corresponds to the mineral matildite, is hexagonal and the high temperature phase α-$AgBiS_2$, which corresponds to the mineral schapbachite, is cubic [17–19]. Experiments on stoichiometric $AgBiS_2$ indicate that the cubic form (α-$AgBiS_2$) is stable above $T = (468 \pm 5)$ K [16, 19] and coexists with the hexagonal form (β-$AgBiS_2$) in the temperature range from 468 ± 5 to 455 ± 3 K, and that below $T = (455 \pm 3)$ K, only the hexagonal form is the stable phase [16].

In the ternary Ag-Cu-S system, three equilibrium ternary phases have been identified: $Ag_{1.55}Cu_{0.45}S$ (Jalpaite, α-phase), $Ag_{1.2}Cu_{0.8}S$ (mckinstryite, β-phase) and $Ag_{1-z}Cu_{1+z}S$ (stromeyerite, γ-phase) [20–23]. All the ternary phases exist on the Ag_2S-Cu_2S tie line and are stable only at temperatures below $T = 392$ K [21, 22]. According to Skinner [20] and Frueh [24], stomeyrite ($Ag_{1-z}Cu_{1+z}S$) exhibits extended stoichiometric variation of $0 < z < 0.1$. Thus, slight differences in the melting temperatures of stromeyerite reported in the literature can be expected. The $Ag_{0.89}Cu_{1.10}S$ reported by Tokuhara et al. [25] as the stable phase of stromeyerite is in agreement with the estimated stoichiometric variation of stromeyerite. In this study, we have synthesized stromeyerite in excess of Cu_2S which has an approximate composition $Ag_{0.93}Cu_{1.07}S$. Stromeyerite with a composition AgCuS has been reported to be synthesized through different methods [26, 27]. At $T = 363$ K, $Ag_{0.93}Cu_{1.07}S$ transforms congruently to the hexagonal solid solution of Cu_2S, i.e. $(Cu, Ag)_2S$ (hcp) [21]. The Cu-rich solid solutions transforms into cubic solid solutions $(Cu, Ag)_2S$ (fcc) [21, 28]. The fcc solid solutions have homogeneity regions on both sides starting from the ideal stoichiometric compositions Cu_2S and Ag_2S [28].

Pyrometallurgical Roasting

The pyrometallurgical roasting of sulfide ores and concentrates is an important process step in the production of metals and chemicals. In this roasting process sulfides of iron, zinc, copper and other valuable metals are oxidized with air in the temperature range from 873 to 1273 K into metal oxides and gaseous sulfur dioxide. After cleaning and cooling, the sulfur dioxide contained in the roasting gas is further processed to produce sulfuric acid. Heat recovery is essential part of the entire roasting processes, both in the complete or partial roasting of sulfides. Sulfide ores and concentrates roasting in the state-of-the-art fluidized bed reactors combine efficient heat recovery and off-gas treatment, including the process of converting the off-gas into sulfuric acid [29]. Heat is recovered from the exothermic processes such as dead roasting of Zn concentrates mainly in the form of high-pressure steam production.

Experimental Section

Materials Preparation

The ternary phase $AgBiS_2$ was synthesized from a mixture of fine powders of 99.9+ % pure Ag_2S and 99.999% pure Bi_2S_3, which were both purchased from

Alfa Aesar (Germany). The pure binary phases in powder form were mixed in the appropriate composition, pressed into a pellet, sealed in an evacuated fused silica tube, and then annealed in a muffle furnace at $T = 673$ K for three days and at $T = 873$ K for twelve days. The synthetic phase was quenched in ice water, while the fused silica glass tube holding it was intact. A sample of the synthesized material was prepared and analyzed by the SEM-EDS techniques. The SEM-EDS analyses confirmed the existence of a single homogenous ternary phase, with an approximate composition of $AgBi_{0.98}S_{1.95}$. X-ray Powder Diffraction (XRPD) analysis of the synthesized sample also confirmed that there were no impurity phases within 0.05 mass fraction XRPD detection limits.

In order to synthesis the ternary phase $Ag_{0.93}Cu_{1.07}S$, appropriate compositions of fine powders of 99.9 + % pure Ag_2S and 99.5% pure Cu_2S were weighed and mechanically well-mixed, in excess of Cu_2S. The mixture of the two binary phases were pressed into pellets, sealed in an evacuated fused silica tube and annealed in a muffle furnace at $T = 573$ K for three days and at $T = 873$ K for ten days. In an attempt to homogenize the sample as well as to make sure traces of Ag_2S may not exist, it was heated up to $T = 1173$ K and then cooled down to $T = 338$ K. The sample was then annealed at 338 K for three weeks. The synthesized material was analyzed by the SEM-EDS technique. The SEM-EDS analyses confirmed the existence of a homogenous ternary phase with a composition of the mineral stromeyerite, $Ag_{0.93}Cu_{1.07}S$, coexisting with excess of Cu_2S.

Thermal Analysis

Thermal analyses of the synthesized samples were done by a simultaneous Differential Thermal Analysis (DTA)- Thermogravimetric Analysis (TGA) analyzer using TA Instruments SDT Q600. The calorimeter was calibrated with the melting temperatures of high purity zinc, aluminum, and gold. The average measurement accuracies of temperatures were determined to be ± 2 K.

Two calorimetric measurements were performed for each phase. Mass change and heat flow during the linear heating and cooling in all runs were measured simultaneously. The runs were performed in a synthetic air with a composition of 20% O_2/80% N_2 gases. In all runs the furnace was heated to 1173 K and then cooled to 573 K. The heating and cooling rates were 5 K min^{-1}, and the flow rate of the gas was 100 ml min^{-1}. Al_2O_3 crucibles were used as a sample holder and reference in all runs. 69.1 mg of fine powders of $Ag_{0.93}Cu_{1.07}S$, and 50.9 mg of fine powders of $AgBiS_2$ were analyzed separately in the different runs.

Results and Discussion

The DTA-TGA curve for the analyzed sample $Ag_{0.93}Cu_{1.07}S$ in air up to 1173 K is given in Fig. 1. An endothermic and two major exothermic peaks can be noticed in the heating DTA curve. Even if the analyzer was not calibrated to determine enthalpy values accurately the relative values associated with each peak can be compared. The first endothermic onset temperature $T = (360.9 \pm 2)$ K corresponds to the phase transition (reaction (1)) temperature determined in an argon gas atmosphere, $T_{tr} = (361.2 \pm 1)$ K [30]. However, the enthalpy value 5.89 kJ mol^{-1} associated with the endothermic peak is higher than the $\Delta_{tr}H^{o} = (4.832 \pm 0.055)$ kJ mol^{-1} value determined for reaction (1) [30], after calorimetric measurements in argon gas atmosphere. In addition to the inaccuracy in the setup of the DTA analyzer employed in this study for enthalpy measurements, the $\sim$1 kJ mol^{-1} difference may also be partly attributed to the different gas atmospheres used in the current measurements.

$$Ag_{0.93}Cu_{1.07}S(\text{ortho}) \rightleftarrows (Cu, Ag)_2S(\text{hcp}). \qquad (1)$$

As shown in Fig. 1, after the endothermic peak, two exothermic peaks accompanied by significant gains in mass have appeared. These are indications for oxidation of the solid solution (Cu, Ag)$_2$S(hcp), which is the stable phase above $T = (361.2 \pm 1)$ K [30]. The SEM-EDS analyses of the oxidized sample, for which the Backscattered Scanning Electron micrographs is shown in Fig. 2, indicate formations of CuO, Ag_2SO_4, and Ag (with 3.88 wt% Cu dissolved). Based on the standard Gibbs energies of reactions (2)–(6) illustrated in Fig. 3, and the DTA-TGA

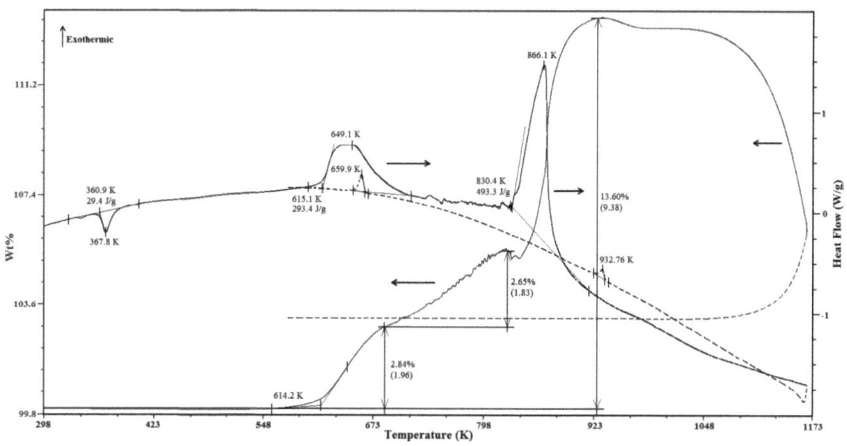

Fig. 1 The DTA-TGA curves of the oxidation processes of the synthesized sample $Ag_{0.93}Cu_{1.07}S$. *Dashed lines* indicate cooling curves

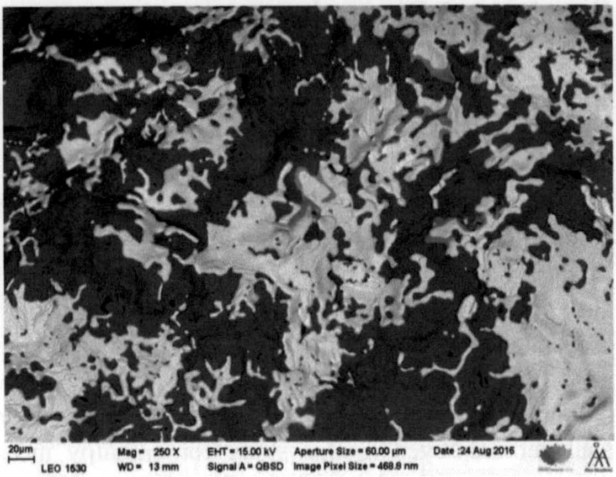

Fig. 2 Backscattered Scanning Electron micrographs of the oxidized $Ag_{0.93}Cu_{1.07}S$ sample after the simultaneous DTA-TGA analyses. They are composed of CuO, Ag_2SO_4, and Ag (with 3.88 wt% Cu dissolved)

curves as a function of temperature shown in Fig. 1, the oxidation process above $T = (614.2 \pm 2)$ K is defined as follows:

$$\begin{gathered}(\mathrm{Cu,Ag})_2\mathrm{S} + \mathrm{O_2(g)} \rightleftarrows \mathrm{CuO} + \mathrm{Ag_2S} + \mathrm{SO_2(g)} \\ (\text{Over all reaction}: \mathrm{Cu_2S} + 2\mathrm{O_2(g)} \rightleftarrows 2\mathrm{CuO} + \mathrm{SO_2(g)}),\end{gathered} \tag{2}$$

$$\mathrm{Ag_2S} + \mathrm{O_2(g)} \rightleftarrows 2\mathrm{Ag} + \mathrm{SO_2(g)}, \tag{3}$$

$$\mathrm{Ag_2S} + 1.5\mathrm{O_2(g)} \rightleftarrows \mathrm{Ag_2O} + \mathrm{SO_2(g)}, \tag{4}$$

$$\mathrm{Ag_2S} + 2\mathrm{O_2(g)} \rightleftarrows \mathrm{Ag_2SO_4}, \tag{5}$$

$$\mathrm{O_2(g)} + 2\mathrm{Ag} + \mathrm{SO_2(g)} \rightleftarrows \mathrm{Ag_2SO_4}. \tag{6}$$

The first exothermic peak above the onset temperature $T = 615.1$ K is due to reaction (2), and the second exothermic peak above the onset temperature $T = 830.4$ K is due to reaction (6). The corresponding standard Gibbs energies for reactions (2)–(6) are calculated with the thermodynamic database of HSC Chemistry 6 [31] and illustrated in Fig. 3.

In Fig. 1 the mass gain above $T = (614.2 \pm 2)$ K is due to the oxidation of Cu_2S and Ag_2S in the solid solution $(Cu, Ag)_2S$ according to reactions (2) and (5), respectively. This temperature of oxidation reaction is in agreement with the peak temperature $T = 605$ K reported by Sceney et al. [32] while oxidizing Cu_2S. The decrease in mass observed above $T = 780$ K is due to the increased loss of sulfur in the form of $SO_2(g)$, according to reaction (3). These observed temperature value and the effect on TGA curve is in agreement with the results reported by Živković et al. [33] for oxidation processes of Ag_2S sample. The second exothermic peak was accompanied with a sharp increase in mass. This sharp increase in mass can be due to the formation of Ag_2SO_4, according to reaction (6), which also has relatively higher enthalpy of reaction. As the Gibbs energies of reactions in Fig. 3 shows, the

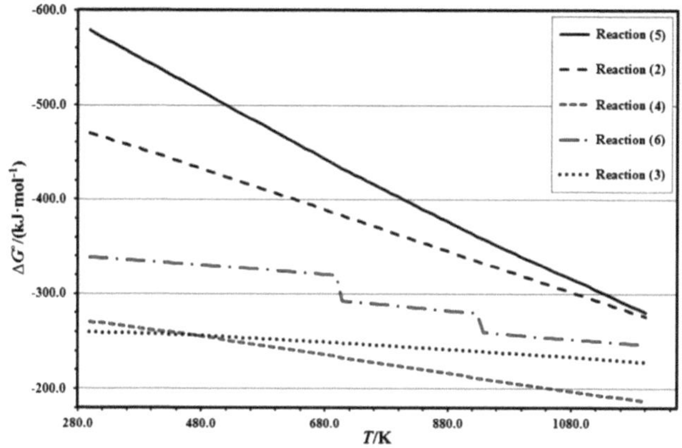

Fig. 3 Standard Gibbs energies of reactions (2)–(6) calculated using the databases of thermodynamical and process calculation software HSC Chemistry 6 [31]

formation of Ag_2O according to reaction (4) is not thermodynamically favored over reaction (3). This is also observed in the analysis of our oxidized sample that contains only silver instead of Ag_2O. This is in agreement with the observation of Živković et al. [33]. The endothermic peak on the cooling curve at $T = (932.76 \pm 2)$ K corresponds well with the melting temperature of Ag_2SO_4, $T = 931$ K, determined by Singh et al. [34]. The endothermic peak on the cooling curve at $T = (659.9 \pm 2)$ K could be due to the phase transition of the produced α-Ag_2SO_4 to β-Ag_2SO_4, which is close to the phase transition temperature determined by Kumari and Secco [34] in a protected atmosphere, $T_{tra} = 689$ K.

An attempt to independently oxidize $AgBiS_2$ sample and measure DTA-TGA values failed due to the reaction of the products with Al_2O_3 crucible. The SEM-EDS analyses of the analyzed sample indicate formations of Al-based phases including Al_2BiO_4, Backscattered Scanning Electron micrographs of the oxidized $AgBiS_2$ sample is shown in Fig. 4. According to the high-temperature calorimetric

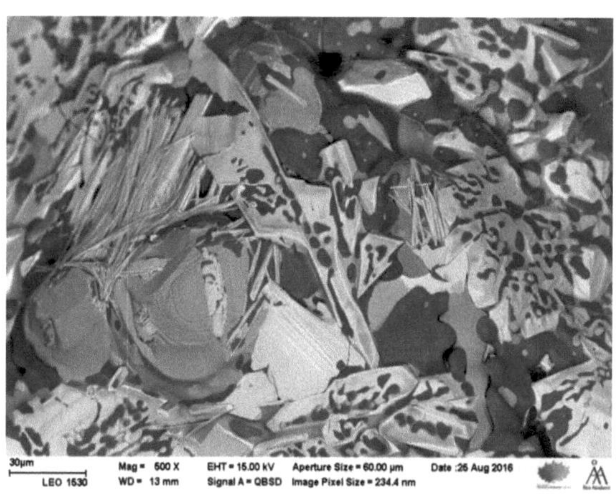

Fig. 4 Backscattered scanning electron micrographs of the oxidized $AgBiS_2$ sample after the simultaneous DTA-TGA analyses. They constitute different alloys and compounds of Al-O-Bi-Ag

studies of Oudich et al. [35], Bi_2O_3 readily reacts with Al_2O_3 in air to form two ternary phases. Thus, the formations of Al–Bi-O phases indicate that Bi_2O_3 has been produced in the high temperature oxidation of $AgBiS_2$.Thus, the possible oxidation process of the sample $AgBiS_2$ could be:

$$2AgBiS_2 + 5.5O_2(g) \rightleftarrows 2Ag + Bi_2O_3 + 4SO_2(g). \quad (7)$$

Summary and Conclusions

Thermal stabilities of $Ag_{0.93}Cu_{1.07}S$ and $AgBiS_2$ in an oxidizing atmosphere have been studied by applying the simultaneous DTA-TGA analyses. Based on the DTA measurements the phase transition of stromeyerite to the solid solution $(Cu, Ag)_2S$ (hcp) in air is determined to be T = (360.9 ± 2) K, which is in agreement with the value T = (361.2 ± 1) K reported after calorimetric measurements in an argon atmosphere [30]. For the first time, maximum thermal stability of $(Ag, Cu)_2S$ in an oxidizing atmosphere close to partial pressure of oxygen in air ($P(O_2) \approx 0.20$ atm) is determined to be below T = (614 ± 2) K, above which it oxidizes to form Ag, CuO and Ag_2SO_4. The melting temperature of Ag_2SO_4 in air determined from the cooling DTA curve in Fig. 1, T = (932.76 ± 2) K, is in good agreement with the melting temperature reported by Singh et al. [34], T = 931 *K*.

An attempt to study the oxidation processes of $AgBiS_2$ failed due to the reaction of the products with Al_2O_3 crucible. However, based on the SEM-EDS analysis of the oxidized sample, it's evident that Bi_2O_3 has been produced, which implies the oxidation process for $AgBiS_2$ may have occurred according to reaction (7). It has been found that for measurements of Bi containing samples at elevated temperatures and in an oxidizing atmosphere Al_2O_3 crucibles are not recommended.

Acknowledgements The authors are grateful to the Academy of Finland for financial support. This work was made under the project "Chemistry of biomass impurities at reducing conditions in future thermal conversion concepts" as part of the activities of the Johan Gadolin Process Chemistry Center at Åbo Akademi University.

References

1. S. Staude, A. Dorn, K. Pfaff, G. Markl, "Assemblages of ag–bi sulfosalts and conditions of their formation: the type locality of schapbachite ($Ag_{0.4}Pb_{0.2}Bi_{0.4}S$) and neighboring mines in the Schwarzwald ore district, Southern Germany," Can. Mineral. 48 (2010) 441–466.
2. J.C. Kopp, V. Spieth, H.-J. Bernhardt, Z. dt. Ges. Geowiss, "Precious metals and selenides mineralisation in the copper-silver deposit Spremberg-Graustein, Niederlausitz, SE-Germany," 163/4 (2012), 361–384.
3. J.R. Craig, G. Kullerud, "The Cu-Zn-S system," Mineral Deposita 8 (1973), 81–91.
4. D. Chen et al., "Microwave synthesis of AgBiS2 dendrites in aqueous solution," Inorg. Chem. Commun. 6 (2003) 710–712.

5. L.K. Samanta, S. Chatterjee, "On the linear, nonlinear, and optoelectronic properties of some multinary compound semiconductors," Phys. State. Sol. (b) 182 (1994), 85–89.
6. T. Thongtem, N. Tipcompor, S. Thongtem, "Characterization of $AgBiS_2$ Nanostructured Flowers Produced by Solvothermal Reaction," Mater. Lett. 64 (2010) 755–758.
7. G.Z. Shen et al., "Novel polyol route to $AgBiS_2$ nanorods," J. Cryst. Growth 252 (2003), 199–201.
8. J.Q. Wang et al., "Synthesis of AgBiS2 microspheres by a templating method and their catalytic polymerization of alkylsilanes," Chem. Commun. 46 (2007), 4931–4933.
9. H. Liu et al., "A mild biomolecule-assisted route for preparation of flower-like $AgBiS_2$ crystals," J. Alloys Compd. 509 (2011), 267–272.
10. N.K. Allouche et al., "Influence of aluminum doping in $CuInS_2$ prepared by spray pyrolysis on different substrates," J. Alloys Compd. 501 (2010) 85–88.
11. M. Lei et al., "Cathodoluminescence variation of a single tapered CdS nanowire," J. Alloys Compd. 509 (2011), 5020–5022.
12. J. Yan et al., "Synthesis of Cu_3BiS_3 and $AgBiS_2$ crystallites with controlled morphology using hypocrellin template and their catalytic role in the polymerization of alkylsilane," J. Mater. Sci. 47 (2012), 4159–4166.
13. M. Trots et al., "High-temperature thermal expansion and structural behavior of stromeyerite, AgCuS," J. Phys.: Condens. Matter 19 (2007), 136–204.
14. R.F. Kadrgulov et al., "Phase relations, ionic transport and diffusion in the alloys of Cu_2S-Ag_2S mixed conductors," Ionics 7 (2001) 156–160.
15. H. Zhu et al., "Room-temperature synthesis of $(Ag,Cu)_2S$ hollow spheres by cation exchange and their optical properties," Mater. Chem. Phys. 127 (2011) 24–27.
16. J.R. Craig, Phase relations and mineral assemblages in the Ag-Bi-Pb-S system. Mineralium Depozita 1(1967), 278–305.
17. S. Geller, J.H. Wernik, Ternary semiconducting compounds with sodium chloride-like structure: $AgSbSe_2$, $AgSbTe_2$, $AgBiS_2$, $AgBiSe_2$. Acta Cryst 12 (1959), 46–54.
18. A.C. Glatz, A. Pinella, X-ray and Neutron Diffraction Studies of the High-Temperature 13-Phase of the $AgBiSe_2$/$AgBiS_2$ System. J Mater Sci 3 (1968), 498–501.
19. D. Wu, The stability of matildite($AgBiS_2$) and $Ag_2Bi_4S_7$ and phase relations in the system Ag_2S-Bi_2S_3. Acta Mineralogica Sinica 9 (1989), 126–132.
20. B.J. Skinner, "The System Cu-Ag-S," Econ. Geol. 61(1966), 1–26.
21. Y.A. Chang, J. P. Neumann, U.V. Choudary, Phase Diagrams and Thermodynamic Properties of Ternary Copper-Sulfur-Metal Systems, INCRA Monograph VII, The Metallurgy of Copper, NBS, Washington, 1979.
22. D. Wu, "Phase Relations in the System Ag_2S-Cu_2S-PbS and Ag_2S-Cu_2S-Bi_2S_3, and Their Mineral Assemblages," Chin. J. Geochem. 6 (1987) 225–233.
23. S. Djurle, "An X-ray study on the system Ag–Cu–S" Acta Chem. Stand. 12 (1958) 1427–1436.
24. A.J. Frueh, "The crystal structure of stromeyerite, AgCuS: A possible defect structure," Z. Kristallogr. 106 (1955) 299–307.
25. Y. Takuhara et al., "Syntheses of complex sulfides AgCuS and Ag3CuS2 from the elements under hydrothermal conditions," J. Ceram. Soc. Jpn. 117 (2009) 359–362.
26. S.N. Guin et al., "Temperature dependent reversible p-n-p type conduction switching with colossal change in thermopower of semiconducting AgCuS," J. Am. Chem. Soc. 136 (2014), 12712–12720.
27. J.A. Schmidt, A.E. Sagua, "Thermodynamic quantities for the ternary compound Stromeyerite: $Cu_{1+\delta}Ag_{1-\delta}S$ for $0 < \delta < 0.1$," J. Chem. Thermodynamics 25 (1993) 1453–1459.
28. R.F. Kadrgulov, R.A. Yakshibaev, M.A. Khasanov, Ionics 7 (2001) 156–60.
29. "Outotec roasting solutions," (Sustainable use of Earth's natural resources, 2016), 8. Available at http://www.outotec.com, Accessed: September 2016.
30. F. Tesfaye, D. Lindberg, P. Taskinen, "Solid state electrochemical and calorimetric study of the equilibrium phase (Cu, $Ag)_2S$," 94 (2016), 101–109.

31. A. Roine et al., "HSC Chemistry 6," Outotec Oy Research Centre, Finland, (2010).
32. C.G. Sceney et al., "Thermal analysis of copper dithiocarbamates," 11 (1975), 301–306.
33. D. Živković et al., "Thermal study and mechanism of Ag_2S oxidation in air," J. Therm. Anal. Calorim. 111 (2013), 1173–1176.
34. K. Singh et al., "Investigation of the Ag_2SO_4- $BaSO_4$ binary system from an SOx sensor point of view," Ionics 8 (2002), 470–478.
35. F. Oudich, et al., "Phase equilibria investigations and thermodynamic modeling of the system Bi_2O_3–Al_2O_3," J. Nucl. Mater. 457 (2015), 72–79.

Oscillation of a Zirconium Droplet—Experiment and Simulations

Kaushal Sumaria, Robert W. Hyers and Jonghyun Lee

Abstract Surface tension is an important property for processing of molten metals. Therefore, it is important to measure the surface tension accurately. Conventional contact methods for surface tension measurements compromise on accuracy due to impurities and oxidation at the molten metal and atmosphere interface. A contactless technique is being utilized for a more accurate measurement of surface tension of high temperature molten metals. A molten zirconium sample is electrostatically levitated using an electrostatic levitator at NASA Marshall Space Flight Center (MSFC) and oscillated at its natural frequency. The surface tension of the sample is related to the frequency of oscillation by Rayleigh's formula. The oscillation of a molten metal droplet was simulated using a CFD package Fluent. Simulated oscillation was analyzed using Matlab code and fast Fourier transform was performed to extract oscillation frequency. Results from the simulations were in good agreement with the experimental data.

Keywords Surface tension · Zirconium · Electrostatic levitation · Computational fluid dynamics

Introduction

Surface tension dominates many metallurgical processes. For example, in steel casting process, high interfacial tension between molten steel and slag prevents drop-off particles in molten steel. Conventionally, sessile drop or pendant drop techniques are used for surface tension measurement. Experiments conducted using these conventional methods contain large scatter between available data due to oxidation at the surface and contamination at the interface with a container [1]. An alternative to these techniques is to determine surface tension by the oscillating drop

K. Sumaria · R.W. Hyers · J. Lee (✉)
Department of Mechanical and Industrial Engineering, University of Massachusetts, Amherst, MA 01002, USA
e-mail: jonghyunlee@ecs.umass.edu

A. Allanore et al. (eds.), *Materials Processing Fundamentals 2017*,
The Minerals, Metals & Materials Series, DOI 10.1007/978-3-319-51580-9_7

Fig. 1 Levitated sample heated by a heating laser [2]

method using (ESL) [2]. As it avoids any contact with a crucible thus eliminating surface contamination, it reduces systemic errors. The restoring force for surface oscillations is the surface tension, which therefore can be related to the frequency of the oscillations [1] (Fig. 1).

ESL is a contactless method for laboratory processing of materials such as steels, super alloys, refractory metals, semiconductors, etc. ESL can be used to measure various thermophysical properties of the sample such as temperature, surface tension, density, viscosity, emissivity, and creep properties [2]. In contactless measurements, a sample is electrically charged by a D_2 arc lamp and levitated in an electrostatic field formed by a pair of vertical electrodes. The position of the sample is controlled by two additional pairs of electrodes in horizontal plane using position sensors and a feedback system. To control the processing temperature, the sample is heated using a Nd:YAG heating laser and the sample temperature is measured using an optical pyrometer. To record the oscillation behavior of the sample during the surface tension measurements, a high temporal resolution camera is used. The captured video data is analyzed to extract the oscillation frequency from which the surface tension can be determined.

Theory

Drop oscillation dynamics is a classical problem in hydrodynamics. The oscillations of the drop surface can be related to the surface tension by the formula [1],

$$\omega = \sqrt{\frac{l(l-1)(l+2)\sigma}{\rho R_o^3}} \ldots \tag{1}$$

where, ω is the frequency of oscillations, l is the mode of oscillation, ρ is the density of the sample, R is the radius of the sample and σ is the surface tension between sample and surrounding atmosphere. However, we are interested only in mode 2 oscillations frequency which is obtained by simplifying Eq. (1). Mode 2 oscillations frequency is simply given by Rayleigh's formula,

$$\omega_0^2 = \frac{8\sigma_0}{\rho_0 R_0^3} \cdots \quad (2)$$

Experiment

A pure zirconium sample was tested using an electrostatic levitator at NASA MSFC. The levitated sample was excited an additional electrostatic field alternating near the natural frequency of the sample. A function generator was used to regulate the excitation frequency. Once enough excitation was achieved, the function generator was turned off and the sample was allowed to damp freely. The high speed video was taken at a frame rate of 1000 fps. The image processing was performed to extract the change in the projected area of the sample during damping as a function of time. An exponential decay function was fitted through the area-time plot to determine the oscillation frequency of the sample.

The experiments were performed in vacuum conditions of 10^{-7} torr. The mass of the sample is 43.288 mg [3]. Mass loss f the sample due to evaporation is as low as 0.043% and the change in mass during processing was neglected. The sample went through one melt cycle to determine the density. Further, the sample went through twenty-four oscillation cycles at six different temperature holds between 1974 and 1600 °C. The change in projected area of each sample was captured using a LED back light and a high speed camera and the captured videos were analyzed to extract the oscillation frequency and thus surface tension.

Numerical Simulation

A numerical model was developed to simulate the oscillation of a pure Zr droplet using the ANSYS Fluent CFD package. For the numerical modeling, one arbitrary oscillation at 1974 °C was used. For this case, the measured oscillation frequency and the corresponding surface tension are 168.88 Hz and 1.452 $N{\cdot}m^{-1}$ respectively. The simulations had two phases, zirconium and helium, with zirconium forming the inner phase and helium forming the surrounding atmosphere. Densities of zirconium and helium used in the numerical model were 5891.74 and 0.1625 $kg{\cdot}m^{-3}$ respectively [4]. The mass loss in sample before and after the experiment is as 0.043% and hence neglected in simulation. Since the experiment is conducted in vacuum, oxidation was neglected in the numerical model. Diameter of the zirconium phase was calculated using density and mass of the sample from the experiments. The viscosity of Zr was set as 4.194 mPa·s [5].

Fig. 2 Geometry of the problem. *Black portion* represents the helium phase and *white portion* is the Zirconium phase

The geometry, mesh and solution parameters are discussed below. A video was generated using images at each timeframe to observe the oscillations. A code was developed using Matlab to analyze the video and was discussed further in this paper. The parameters for the numerical model are discussed below.

Geometry

To study the oscillation of a levitated zirconium drop in helium atmosphere, a 2D axisymmetric model is created as shown in Fig. 2. A rectangular domain (4 mm × 2 mm) which includes both zirconium and helium was created. On the left edge of the rectangular, a semicircle with diameter of 2.5 mm was created to define the zirconium drop.

Numerical Solver

The pressure-based solver is used for the numerical simulation. The reason for this is the use of volume of fluid model for multiphase simulations discussed later. Also, a pressure-based solver is used as the fluid can be assumed to be incompressible. The pressure based solver satisfies mass conservation by solving the pressure equation. The transient time solver is implemented and simulation is carried out until the desired time-length of data is obtained to analyze and extract frequency from it.

Multiphase Modelling

Volume of fluid model consisting of two phases is used for the simulations. Volume of fluid method is used for problems where the multiple phases do not mix and form a defined boundary at all time. Volume of fluid method solves single momentum equation for all the phases. A new variable whose value is between 0 and 1 is introduced which traces the volume fraction of all the phases in each control volume. The properties of the individual phases at a particular time step are function of volume fraction variable and global property at given time step. Volume fraction parameters were solved using explicit scheme.

Solution Methods

PISO scheme with neighborhood correction factor of 1 is used for pressure-velocity coupling. Spatial discretization schemes are selected suitable for the Volume of fluid method. The gradient is calculated using least squares cell based method. This algorithm is the most accurate algorithm available with competent computation time. Pressure and momentum equations are solved using the PRESTO! And QUICK algorithms respectively. Geo-Reconstruct scheme is used for the volume fraction calculations. This scheme calculates the volume fraction of the phases in each cell and regenerates it using a straight line with slope in each cell. The scheme is very useful in problems where the interface among phases is to be traced accurately. A very smooth interface is generated using Geo-Reconstruct scheme. For the transient formulation, non-iterative time advancement was selected which means all the fluid property equations were solved once per time step.

Mesh

An Eulerian approach is used in the simulation. In the Eulerian approach, rather than following all the particles in a fluid motion, fixed number of points in space are selected and properties of fluid at a particular time step are recorded. In the Eulerian approach, the fluid properties are a function of space and time. The rectangular domain was divided into a grid mesh with each computational cell of equal volume. The mesh size plays an important part in the computation time of the simulation. Considering the accuracy of solution and computation time, a mesh was generated such that each control volume was square with side of 25 microns.

Solution Initialization and Initial Condition

The geometry of the problem consists only of a rectangular domain. The domain is meshed into a grid with each cell size of 25 microns. The problem needs to be initialized and phase volume needs to be assigned to each cell based on the geometry of the problem. A user defined function was developed to initialize the problem and assign the appropriate phases to each cell in the domain. In order to initiate the oscillations, we introduce ellipticity of 5% in the semicircle diameter. All cells within the semicircle were initialized with volume fraction 1 for zirconium phase. Cells outside the semicircle were assigned volume fraction 1 for the helium phase. The volume fraction of cells at interface was generated using the Geo-Reconstruct scheme.

Time Step

The time step for transient simulation is selected considering computation time and divergence. The simulation time needs to be long enough to capture 2 complete oscillations cycles to extract frequency of oscillation. Another important aspect to be considered is courant number. Courant number needs to be controlled for the simulation to proceed and not crash it due to divergence. Therefore, in finer mesh, time step should be selected carefully. In this simulation the time step is selected to be 2e-5 s (Figs. 3 and 4).

Fig. 3 Meshed domain with cell size of 25 microns

Fig. 4 Initial condition with 5% ellipticity

Results and Discussion

The simulation was initialized and run for simulation time of 0.01763 s using 1764 time steps. The simulation was recorded and compiled into a video using the feature of Ansys post processing package. The image at each time step was recorded and acted as one frame in the video. The video was recorded at the speed of 30 frames per second. Images at different time-step during one oscillations are shown below.

In order to get the frequency of oscillations, the area of zirconium phase is to be recorded and plotted against corresponding time step. A code is developed using Matlab. First the video is read into the memory and each frame is analyzed. Each frame is then converted to greyscale image. Each pixel of the image is then analyzed using a nested loop and the greyscale value at each cell is stored. Number of cells having the value of more than 29 is added to a counter variable and is stored as area of zirconium phase for that time step. Subsequently, all frames are analyzed and plotted to obtain the oscillation spectrum.

A fast Fourier transform is performed on the data set and a frequency spectrum is generated. A sharp peak is obtained at 170.26 Hz which is the frequency of zirconium sample oscillation in helium atmosphere. The result obtained from numerical simulation are in good agreement with the experiments performed. Further, Eq. 1 can be used to calculate the interfacial or surface tension of Zirconium. Values of density, radius and frequency is put in the equation to obtain the value of surface tension to be 1.473 N/m (Figs. 5, 6 and 7).

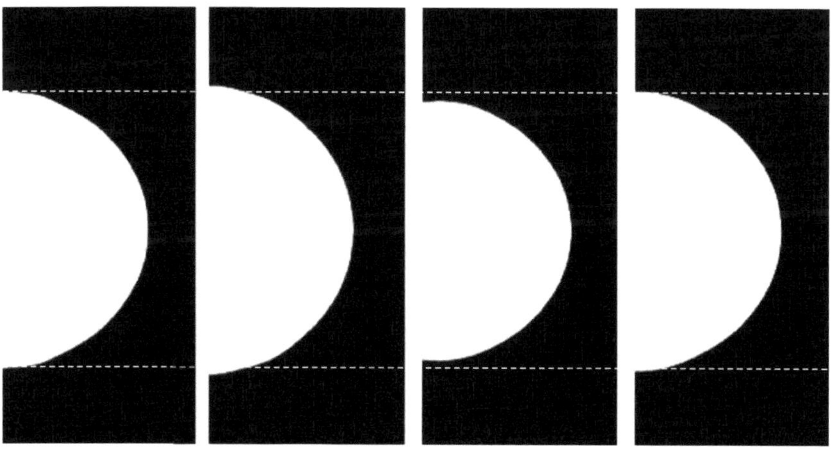

(a) (b) (c) (d)

Fig. 5 Different stages of oscillation. Two dashed lines were added to show the relative difference in the position of zirconium – helium interface. **a** Initial condition with 5% ellipticity, **b** maximum area during oscillation, **c** minimum area during oscillation and **d** mean area during the oscillation

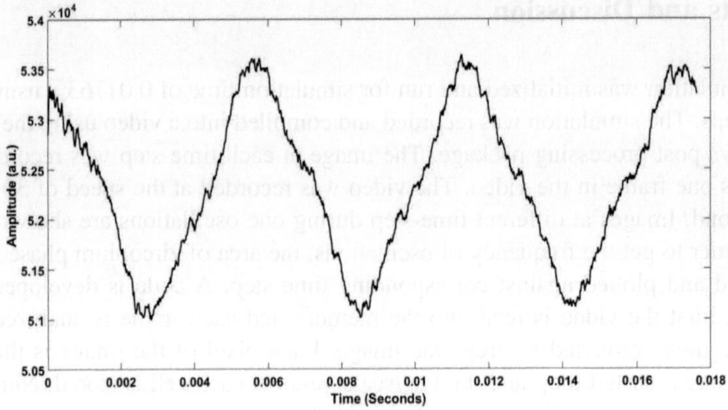

Fig. 6 Oscillation spectrum of Zirconium in helium atmosphere

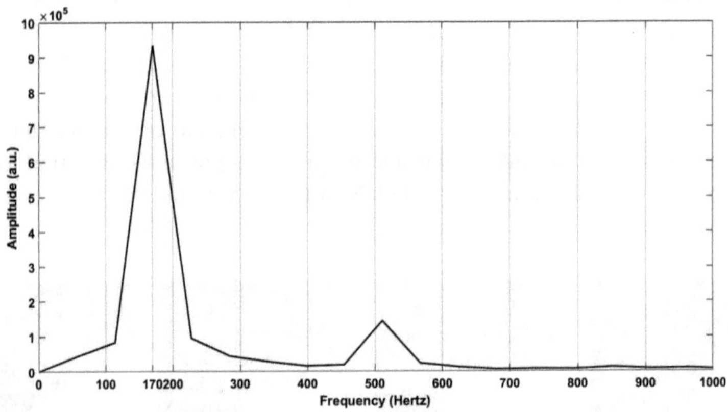

Fig. 7 Frequency spectrum of Zirconium in helium atmosphere

References

1. Egry, I., *Surface tension measurements of liquid metals by the oscillating drop technique.* Journal of Materials Science, 1991. **26**: p. 2997–3003.
2. Robert W. Hyers, J.R.R., *A review of Electrostatic Levitation for Materials Research.* High Temperature Materials and Processes, 2008. **27**(6): p. 461–474.
3. Jie Zhao, J.L., Rainer Wunderlich, Hans Fecht, Stephan Schneider, Michael SanSoucie, Jan Rogers, Robert Hyers, *Influence of Oxygen on surface tension of Zr.* 2016.
4. Takamachi Iida, R.I.L.G., *The physical properties of liquid metals.* 1993, Oxford University Press Inc., New York.
5. Paul-Francois Paradis, W.-K.R., *Thermophysical properties of zirconium at high temperature.* Journal of Materials Research, 1999. **14**(9): p. 3713–3719.

Gallium Evaporation Behavior for Purification in Molecular Beam Epitaxy

Kyungjean Min, David Johnson and Kevin Trumble

Abstract High mobility GaAs/AlGaAs heterostructure have been achieved at Purdue University through in-situ distillation of the source Ga prior to growth. A significant amount of Ga is lost during distillation and growth. To evaluate the Ga behavior during MBE operation, the evaporation rate from a planar source of liquid Ga in a crucible is analyzed, correcting for transmission probability according to Clausing theory. The transmission probability which depends on the diameter and receding depth of the liquid surface in the crucible, explains condensation of Ga particles by collision for the vertically standing crucible. The good agreement furthermore suggests that the usual tilt of the cell has little effect on the flux.

Keywords Gallium purity · Molecular beam epitaxy (MBE) · Langmuir-Knudsen equation

Introduction

Molecular Beam Epitaxy (MBE) is state-of-the-art deposition technique that produces high quality two dimensional electron gas (2DEGs) with high electron mobility by using modulated doping methods from separated elemental source effusion cells under ultra-high vacuum conditions. At Purdue University, GaAs/AlGaAs heterostructures have been grown by MBE with an electron mobility greater than 35×10^6 cm^2/Vs when using a Ga source with an initial purity of 8N (99.999999%) Ga, but a lower electron mobility of 25×10^6 cm^2/Vs was obtained when a 7N (99.99999%) Ga source was used [1, 2].

The ultra-high electron mobility attained at Purdue MBE was recently shown to depend mainly on the purity of Ga source [3], as well as the uniformity of molecular

K. Min (✉) · D. Johnson · K. Trumble
School of Materials Engineering, Purdue University, West Lafayette, IN 47907, USA
e-mail: kjmin@purdue.edu

A. Allanore et al. (eds.), *Materials Processing Fundamentals 2017*,
The Minerals, Metals & Materials Series, DOI 10.1007/978-3-319-51580-9_8

beam pattern associated with angular flux distribution of the source in MBE [4]. Therefore, the investigation of Ga evaporation behavior in the Purdue MBE geometry can contribute to the understanding of correlation between Ga evaporation during in-situ distillation for the geometry used in the MBE is of interest.

From the principle of kinetic theory, Langmuir derived the equation of maximum rate of evaporation from free solid surfaces. Langmuir calculated evaporation rate and vapor pressure of tungsten from filaments in vacuum for a range of temperatures and compared with experimental results. The calculated evaporation rate of tungsten measured in a vacuum agreed with the experimental results [5].

Knudsen developed the evaporation theory for the isothermal effusion cell with small orifice. Knudsen employed the coefficient that was multiplied to the Langmuir equation and pressure of vacuum reservoir that was subtracted from the Langmuir equation. Those terms give an effect of condensation of evaporated particles to surface [6].

Clausing added the condensation factor due to collision between particles inside a crucible to the Knudsen equation and derived the flux of angular distribution. Clausing divided the evaporation behavior of particles into four different cases under the consideration of collision and derived the probability factor including those four cases [7].

More recently, Krasuski calculated the flux of angular distribution on the exit of crucible that has a cylindrical shape by using Monte Carlo simulations based on Clausing's derivation. Krasuski found the flux has significant variation depending on different angles [8].

For the growth of the GaAs/AlGaAs heterostructures described above, the source Ga was distilled within the MBE chamber for further purification. During distillation, and even during growth, a significant amount of Ga is lost. In addition, the evaporated amount of Ga in distillation could not be measured due to the characteristics of MBE experiment to be perfectly evacuated and blocked from outside during operation. By estimating the evaporation amount by thermochemical analysis during distillation and growth, the behavior of Ga during MBE operation and the lost amount of Ga will be numerically analyzed.

Background

The distillation and growth experiments analyzed in this paper were detailed in reference [2]. Ingots of 8N Ga totaling 100 g were loaded to the crucible, which was subjected to insitu distillation for further purification at 1360 K for 16 h. Then 138 heterostructure samples were grown at 1160 K for a total of 400 h. After distillation and growth were finished, it was observed that 25% (25 g) of the initial Ga amount was lost from the effusion cell.

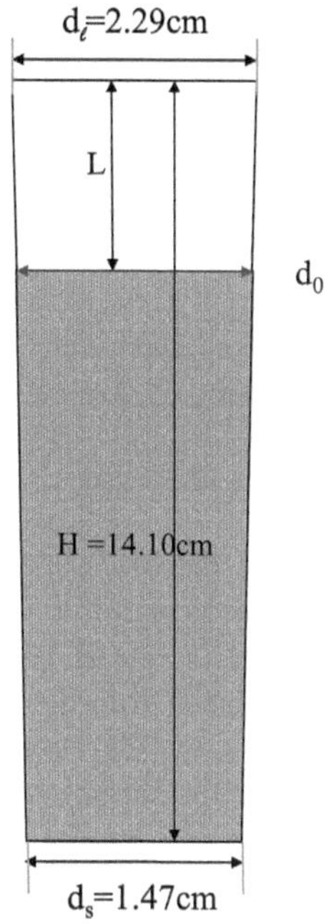

Fig. 1 MBE Ga conical (near-cylindrical) crucible (L_0 = 1.04 cm, d_0 = 1.90 cm)

The Ga crucible is slightly conical, nearly cylindrical. The dimensions of the crucible and the estimated values of diameter and initial depth of liquid, L, when the crucible is filled with 100 g liquid Ga are illustrated in Fig. 1.

The evaporated Ga amount was calculated by Langmuir-Knudsen equation and corrected with the transmission probability of Clausing Theory.

Results and Discussion

The heterostructure growth was performed in ultra-high vacuum of about 1×10^{-10} Torr (=mmHg = 7.6×10^{-3} Pa). For such high vacuum, the rate of distillation is controlled by the rate of evaporation of atoms from the surface of the melt and equilibrium is not maintained between the evaporation from liquid phase and the condensation from vapor phase. This distillation process under the high vacuum condition is called molecular distillation. When there is no returning condensation from vapor phase to liquid phase, the evaporation rate is maximum. The maximum evaporation rate is given by Langmuir equation written in SI unit as,

$$\mathrm{w}\left(\frac{\mathrm{kg}}{m^2 \cdot s}\right) = 4.37 \times 10^{-3} \cdot \mathrm{P}\sqrt{\frac{M}{T}} \tag{1}$$

Here P is the saturation vapor pressure in Pascal, T is absolute temperature in K, and M is molecular or atomic weight in g/mol.

Free Evaporation

The vapor pressure which corresponds to each temperature from 600 to 1200 °C was derived from [9, 10]

$$\mathrm{Log(P/atm)} = -13984/\mathrm{T} + 6.754 - 0.3413 \times \mathrm{Log(T)} \quad (2)$$

As discussed in the above, the maximum evaporation rate under the perfect vacuum condition is defined by the Langmuir Eq. (1).

For the initial 100 g of Ga in the crucible, the radius of circular liquid surface calculated from the taper of the crucible shown in Fig. 1 is 0.948 cm and surface area of the Ga melt is 2.82 cm^2. As the crucible is tilted to 26° vertically in the MBE, the surface area of Ga melts increases to 3.14 cm^2. Using these values, the evaporation rate of Ga from 800 to 1400 K is shown in Fig. 2. The amount of Ga evaporated can be obtained from Langmuir equation by multiplying surface area of Ga melt that faces with the vapor phase and evaporation time (16 h for distillation and 400 h for growth). These are listed in Table 1 for the distillation and growth steps.

As shown in Table 1, a total of 67.9 g of Ga was lost with the majority being during the distillation step. This is because Ga saturation vapor pressure in distillation is two orders higher than that in growth. Clearly, free evaporation using the Langmuir equation overestimates the Ga loss.

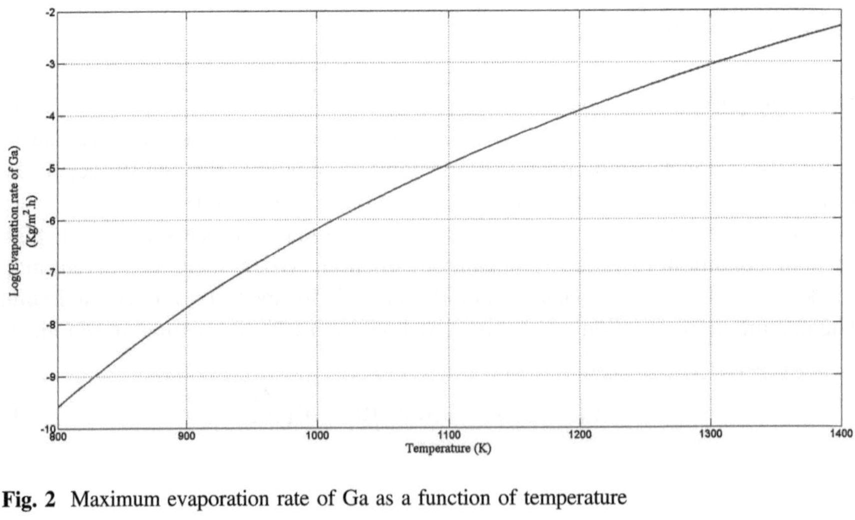

Fig. 2 Maximum evaporation rate of Ga as a function of temperature

Table 1 Maximum evaporation rate and evaporated amount of Ga for distillation and growth

	Time (h)	Temperature (K)	Ga saturation vapor pressure (atm)	Evaporation rate ($kg/m^2 \cdot h$)	Evaporated Ga amount (g)
Distillation	16	1360	4.50E-05	9.12E-02	45.8
Growth	400	1160	2.52E-07	1.76E-03	22.1

Calculation of Ga Gas Particles Transmission Probability for Vertical Standing Crucible

The evaporated particles inside the crucible either pass through without striking the crucible walls or collide with the walls. The molecules that once collided with the walls go through one of the following processes; passing through the exit, returning to the entrance and condensing, or striking with the wall again.

Clausing derived the transmission probability for emitted gas particles from the crucible for the case of particles directly passing through the crucible without collision, and for the case of passing through the crucible with collisions [7].

The motion of gas particles that would evaporate inside the crucible can be divided into the following ways as depicted in Fig. 3: (a) movement from the liquid surface to the crucible opening end, (b) movement from the liquid surface to the wall surface, (c) movement from the wall surface to the crucible opening end, (d) movement from the wall surface to the upper wall surface, and (e) movement from the wall surface to the liquid surface. At low angles, φ, the gas particle evaporates in a straight line without collision. At larger angles, depending on the location of evaporation from the liquid surface, the particle will strike the wall. From the wall there are three possible further processes that may consist of returning back to liquid surface (e), colliding again with the wall surface (d), or direct evaporation to outside of crucible (c).

The transmission probability is defined by Eq. (3).

$$K_c = \int_0^{\mathrm{L}} \mathrm{W_{sr}(x) \cdot w(x)dx + W_{ss}(L)} \tag{3}$$

where W_{sr} is the probability that gas particles move from the liquid surface to the wall surface and W_{ss} is the probability that gas particles move from liquid surface to the crucible opening end. Here, w(x) is the probability of collision at x inside the crucible and is defined by Eq. (4)

$$\mathrm{w(x)} = \int_0^L \mathrm{W_{rr}}(\xi - x)d\xi \cdot \mathrm{w}(\xi) + \mathrm{W_{rs}}(L - x) \tag{4}$$

where ξ is arbitrary point that collision occurs inside the crucible after the collision at x (Fig. 3), W_{rr} is the probability that gas particles move from the wall surface to the upper wall surface, and W_{rs} is the probability that gas particles move from the

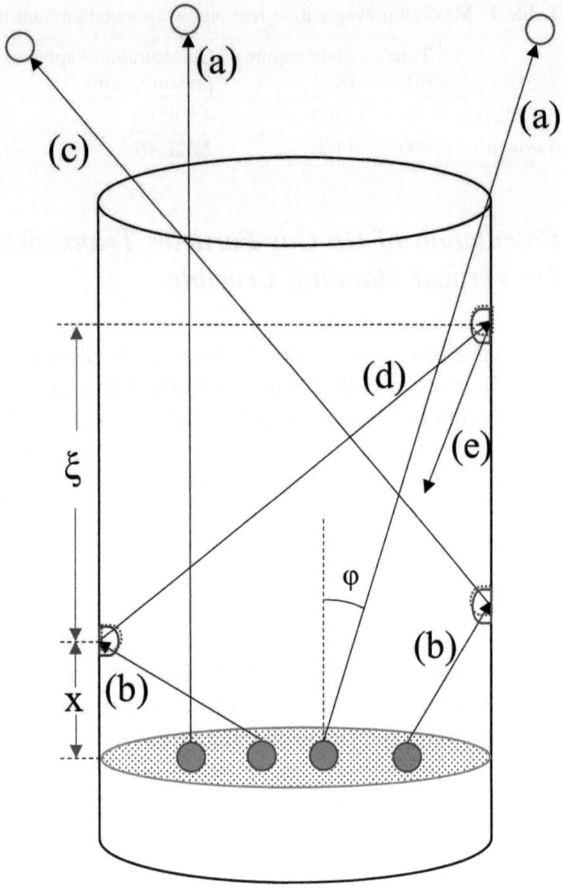

Fig. 3 Schematic diagram of the movement of gas particles inside the crucible in four different cases **a** liquid surface—crucible opening end, **b** liquid surface—wall surface, **c** wall surface—crucible opening end, **d** wall surface—wall surface and **e** wall surface—liquid surface

wall surface to the crucible opening end. The possibility of particles returning to the liquid surface from the wall (case (e) in Fig. 3) is neglected in the Clausing calculation. The probability of collision at x, w(x) can be represented by Eq. (5)

$$\mathrm{w(x)} = \alpha + \frac{1-2\alpha}{L} \cdot x \tag{5}$$

When the distance between melt surface and crucible opening end, L is greater than the radius of melt surface, α is given by Eq. (6)

$$\alpha = \frac{\left[u(u^2+1)^{\frac{1}{2}} - u^2\right] - \left[v(v^2+1)^{\frac{1}{2}} - v^2\right]}{\frac{u(2v^2+1)-v}{(v^2+1)^{\frac{1}{2}}} - \frac{v(2u^2+1)-u}{(u^2+1)^{\frac{1}{2}}}} \tag{6}$$

with $v = \frac{L\sqrt{7}}{3L+2r\sqrt{7}}$ and $u = \frac{L}{2r} - v$.

Then the final transmission probability is represented by Eq. (7)

$$K_c = \frac{1-2\alpha}{3r^2L}\left\{4r^3 + \left(L^2 - 2r^2\right)\left(L^2 + 4r^2\right)^{\frac{1}{2}} - L^3\right\} + \alpha + \frac{1-\alpha}{2r^2}\left\{L^2 - L\left(L^2 + 4r^2\right)^{\frac{1}{2}} + 2r^2\right\} \tag{7}$$

The transmission probability of Ga particles, considering the probability of collision for the MBE crucible is the function of the distance between melts surface and crucible opening end in Fig. 4 and the radius of crucible in Fig. 5 As L is deeper, more collisions with the crucible walls occur. Hence, the transmission probability decreases as L increases in Fig. 4 On the other hand, the dependence of radius is smaller in transmission probability as suggested in Fig. 5.

The evaporation rate at the presence of collision is defined by multiplying transmission probability in Eq. (6) to free evaporation rate in Eq. (1) and given by Eq. (8)

$$\mathrm{W}\left(\frac{\mathrm{kg}}{m^2 \cdot s}\right) = 4.37 \times 10^{-3} \cdot K_c \cdot \mathrm{P}\sqrt{\frac{M}{T}} \tag{8}$$

The evaporation rate is the evaporated weight per unit area of melt surface and unit time. The weight of evaporated amount can be converted to volume change during evaporation. The change of volume in the cylindrical crucible is surface area of melts times change of length of melts. If the surface area of melts is assumed as a constant in cylindrical crucible, the evaporation rate can be represented as a function of receding length in melts during evaporation as indicated in Eq. (9).

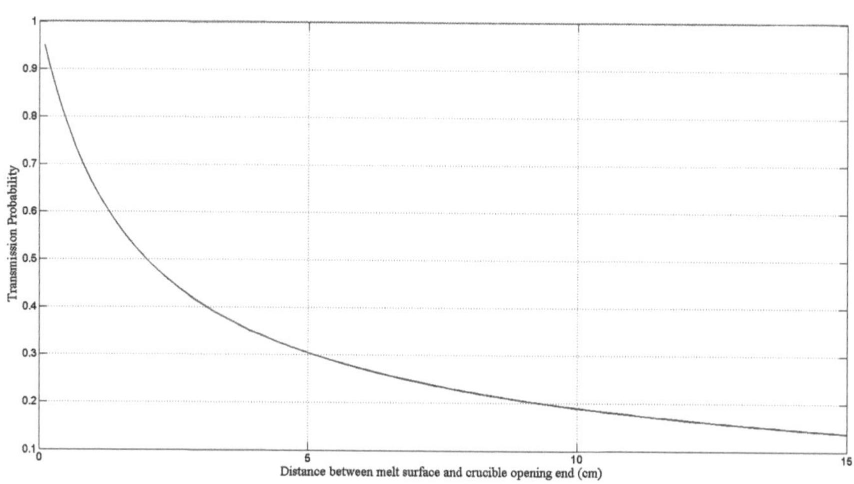

Fig. 4 The variation of transmission probability at the presence of collision in the crucible depending on length between melt surface and crucible end at r = 0.95 cm

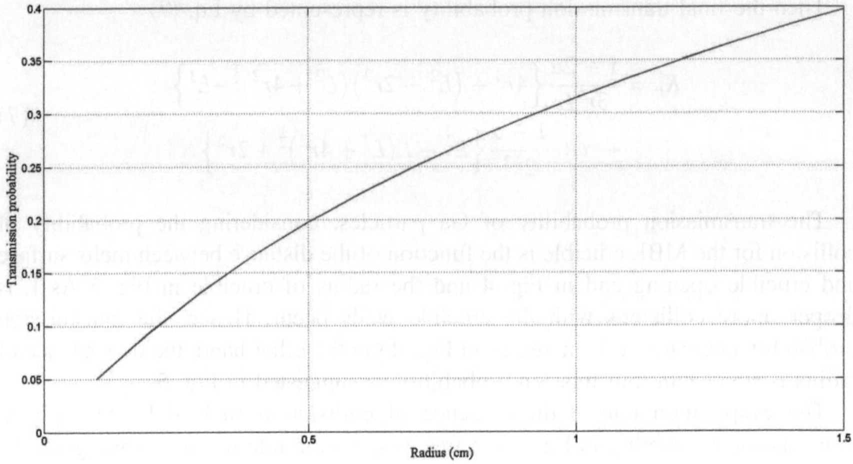

Fig. 5 The variation of transmission probability at the presence of collision in the crucible depending on length between melt surface and crucible end at L = 5 cm

Then the proportion of evaporation amount can be derived by integrating the infinitesimal change of length of melt over time.

$$\begin{aligned} &W(\Delta\,\text{weight}) \rightarrow W(\Delta\,\text{volume}\cdot\text{density}) \\ &\quad\rightarrow W(\Delta\,\text{receding length}\cdot\text{surface area}\cdot\text{density}) \end{aligned} \tag{9}$$

Evaporated Ga Amount Under the Consideration of Collision Inside the Crucible

The evaporated Ga amount with particle collisions can be obtained from the integration of the evaporation rate W in Eq. (8), with the function of receding length, over time. To integrate the evaporation rate for the infinitesimal change of length over time, the equation of receding length that is associated with time is necessary. However, K_c depends on the receding length that varies over time and there is no explicit solution for the integration of W over time. Therefore, the equation between the receding length L and time was derived numerically by the fitting data for each 1 h step of evaporation.

The fitting line for distillation for 16 h is, $L = -0.0002t^3 + 0.0065t^2 - 0.078t$, from the initial parameters of r = 0.95 cm, T = 1360 K and P = 4.50E-05 atm. The fitting line for growth for 400 h is, $L = 10^{-9}t^2 - 2 \times 10^{-6}t + 0.006$, from the initial parameters of r = 0.84 cm, T = 1160 K and P = 2.52E-07 atm. The equations of receding length for distillation and growth are shown in Figs. 6 and 7.

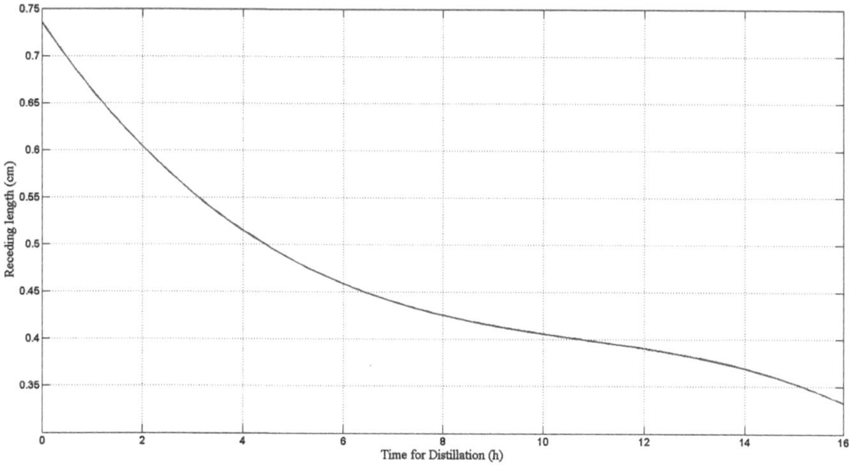

Fig. 6 Receding length for 16 h distillation as a function of time

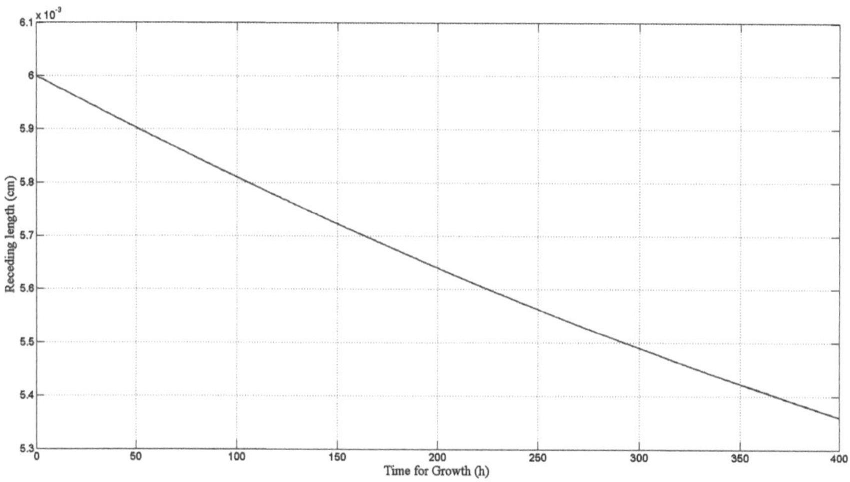

Fig. 7 Receding length for 400 h growth as a function of time

The amount of evaporated Ga for both distillation and growth at whit the presence of collisions are summarized in Table 2 with the comparison to the evaporated amount at free evaporation. Since the crucible was tilted from vertical about 26°, the surface are for derivation of amount was projected to 3.14 cm^2 from 2.82 $cm^{2.}$

The estimated result of 28.9 g was very close to the observation of 25 g. The difference of 3.9 g is because the consideration regarding the 26° tilt of Ga source crucible was not applied. The particles on the 90% of melt surface more probably

Table 2 Maximum evaporation rate and evaporated amount of Ga for distillation and growth

	Time (h)	Temperature (K)	Ga saturation vapor pressure (atm)	Evaporation rate (kg/cm^2 • h)	Evaporated Ga without collision (g)	Evaporated Ga with collision (g)
Distillation	16	1360	4.50E-05	9.12E-02	45.8	23.2
Growth	400	1160	2.52E-07	1.76E-03	22.1	5.7
Sum					67.9	28.9

collide with the crucible wall and condense back to the melt surface in the tilted crucible. Therefore, smaller amount of Ga was evaporated in the observation than calculation. The degree of contribution by condensation in tilted port to the evaporation amount will be discussed in a follow-up paper.

Conclusion

The calculated maximum Ga amount using the Langmuir equation during distillation at 1360 K and growth at 1160 K was 67.9 g. This calculated value was much higher than 25 g of Ga loss observed. This disagreement was corrected by using the transmission probability by Clausing theory which includes the consideration of the condensation of Ga particles back into melt due to collision between Ga particles inside the crucible. The evaporated amount of Ga assuming the presence of collisions was obtained by applying a receding length of melts over time to the evaporation rate within a cylindrical crucible. Using the correction, the evaporated amount of Ga, 28.9 g was closed to the observed amount of loss, 25 g.

Acknowledgements Support for this research from the W.M. Keck Foundation is gratefully acknowledged. The authors also would like to thank ultra-pure GaAs Keck Project team members at Purdue University.

References

1. M. J. Manfra, "Molecular Beam Epitaxy of Ultra-High-Quality AlGaAs/GaAs Heterostructures: Enabling Physics in Low-Dimensional Electronic Systems," Annu. Rev. Condens. Matter Phys, 5 (2014), 347–73
2. G. C. Gardner et al., "Modified MBE Hardware and Techniques and Role of Gallium Purity for Attainment of Two Dimensional Electron Gas Mobility > 35 × 10^6 cm^2/Vs in AlGaAs/GaAs Quantum Wells Grown by MBE," Journal of Crystal Growth, 441 (2016), 71–77
3. Kyungjean Min, "Analysis of High-Purity of Ga by ICP-MS," (MS thesis, Purdue University, 2014)

4. M. A. Herman and H. Sitter, “Molecular Beam Epitaxy, Fundamentals and Current Status,” Springer (1996)
5. I. Langmuir, “The Vapor Pressure of Metallic Tungsten,” Phys. Rev., 2 (11) (1913), 329–342
6. M. Knudsen, “Der molekulare Gaswiderstand gegen eine sich bewegende Platte,” Ann. Phys. 351(5) (1915), 641–656
7. P. Clausing, “Über die Strömung sehr verdünnter Gase durch Röhren von beliebiger Länge,” Ann. Phys. 404(8) (1932), 961–989
8. P. Krasuski, “Angular distribution of flux at the exit of cylindrical tubes,” J. Vac. Sci. Technol. A 5, (1987), 2488–2492
9. C. B. Alcock, VP. Itkin, MK. Horrigan, “Vapour pressure equations for the metallic elements: 298–2500 K,” 23[3] (1984) 309–313
10. O. Kubaschewski and CB. Alcock, “Metallurgical Thermochemistry, 5th ed,” Pergamon Press, (1979)

Mass Transfer of Al and Ca Between Silicon and Synthetic SiO_2-CaO-Al_2O_3 Slags

Erlend Lunnan Bjørnstad and Gabriella Tranell

Abstract The mass transfer of *Al* and *Ca* between silicon and synthetic SiO_2-CaO-Al_2O_3 slags of varying composition has been investigated, using different experimental set-ups. Samples were kept at 1600 °C, varying the holding times in an individual crucible kinetic series set-up or sampled at given sample intervals in a larger crucible set-up. Solute concentrations of *Al* and *Ca* in *Si* and slags were found by ICP-MS. This work compares these results to observations from industrial Si refining measurement campaigns. A model for interpreting kinetic mass transfer phenomena observed for parallel *Al* and *Ca* mass transfer between slag and *Si*, is also discussed.

Keywords Mass transfer coefficient · Silicon production · Refining · Slag mass transfer

Introduction

Metallurgical grade silicon (>96% *Si*) [1] is used as a raw material in aluminum alloy production, silicones, the electronics industry and in renewable energy technologies. The demand for products with a high degree of accuracy, with respect to the composition of impurities, has increased drastically over the last decade. An example of this is the high purity product needed to supply the manufacturing of solar energy technologies. High purity is of utmost importance since even low concentrations of impurities could harm the semiconductor properties of silicon, thus limiting its use in solar panels. To better control the composition of the final product refining steps are added to the production process. Understanding the

E.L. Bjørnstad (✉) · G. Tranell
Department of Materials Science and Engineering, Norwegian University of Science and Technology, Alfred Getz vei 2, 7491 Trondheim, Norway
e-mail: erlend.l.bjornstad@ntnu.no; erlenbjo@stud.ntnu.no

G. Tranell
e-mail: gabriella.tranell@ntnu.no

A. Allanore et al. (eds.), *Materials Processing Fundamentals 2017*,
The Minerals, Metals & Materials Series, DOI 10.1007/978-3-319-51580-9_9

refining process is vital to stay competitive and continue to meet the demands of today, but also the demand of tomorrow.

The typical industrial metallurgical production route for silicon consists of carbothermic reduction of quartz in an electric arc furnace. A majority of the impurities including *Al*, *Ca*, *P*, *B* and *Fe* enter the system through the carbon sources, like coal, charcoal and wood chips. To control the level of impurities in the final product the melt is tapped from the furnace into a ladle to undergo oxidative ladle refining. Oxidative ladle refining is conducted by continuously purging the ladle with oxygen enriched air. After a series of reactions the main impurities, *Ca* and *Al*, will go to an immiscible oxide phase, a slag, which later can be separated from the melt. Understanding the transport and kinetic properties of this process is crucial to exert a more precise control over the products final composition.

Extensive work has already been performed mapping the thermodynamics of the system, however there has been comparatively few studies investigating the mass transfer and reaction kinetics. This work aims to investigate the mass transfer of *Al* and *Ca* between *Si* and SiO_2-CaO-Al_2O_3 slags. Two experiments were performed, but only one yielded publishable results. It also looks at a basic refining model [2] and compares it to the experimental data found by this work, and the industrial data from Kero et al. [3].

Process Overview

As the melt is tapped from the furnace into the ladle the refining process is performed continuously, and continues for a set amount of time after the ladle is full. The purge gas is introduced through a porous plug located at the bottom of the ladle. This creates a bubble column rising through the ladle, driving a flow while mixing the melt. On its ascent the O_2 in the bubbles will react with *Si* to form SiO_2 and $SiO_{(g)}$. $SiO_{(g)}$ will be transported into the bubble saturating the gas phase, while SiO_2 will nucleate at the bubble interface. Tang [4] writes that SiO_2 forms a droplet on the surface, while Schei et al. [5] writes that it can form a film around the bubble. SiO_2 then reacts with the impurities in the melt, primarily *Al* and *Ca*, to form a predominantly SiO_2-CaO-Al_2O_3 slag. When the slag is formed there will be a driving force to approach equilibrium between the present phases driving a mass transport to the slag so their respective equilibrium concentration can be reached. Since the temperature of the system is high it is often assumed that the reactions can be considered instantaneous [6]. As the diffusion of species in the metal is faster than in the slag, the rate determining step for the transport of *Al* and *Ca* is considered to be the transport in the slag [3, 4, 6]. To sum up the overall reactions can be expressed as:

1. $Si + O_2 = SiO_2$
2. $Si + \frac{1}{2}O_2 = SiO_{(g)}$
3. $2\underline{Ca} + SiO_2 = 2CaO + Si$

Fig. 1 The whole system including the falling jet. Bubbles of entrapped air are displayed in *green*

4. $4\underline{Al} + 3SiO_2 = 3Si + 2Al_2O_3$
5. $3CaO + 2\underline{Al} = 3\underline{Ca} + Al_2O_3$

with Fig. 1 illustrating the whole process.

To express the system it has been proposed to use a standard [2] batch reactor model [3]. By applying two-film theory to the interface the molar transfer for species i, N_i, from the metal to slag can be expressed:

$$N_i|_{m\rightarrow s} = \frac{k_{i,m}\rho_m}{M_i}(\omega_{i,m} - \omega_{i,m}^{\star}) = \frac{k_{i,s}\rho_s}{M_i}(\omega_{i,s}^{\star} - \omega_{i,s})\left[\frac{kmol}{m^2 s}\right] \quad (1)$$

where $k_{i,p}\left[\frac{m}{s}\right]$ is the mass transfer coefficient for species i in phase p, $\rho_p\left[\frac{kg}{m^3}\right]$ is the density of phase p, and $M_i\left[\frac{kg}{kmol}\right]$ denotes the molar mass of species i. $\omega_{i,p}$ denotes here the mass fraction of species i in phase p, and the superscript * expresses that it is at the interface. Due to the complexity of determining the mass transfer coefficient, k_i, $k_{i,m}$ and $k_{i,s}$ may be combined in an overall mass transfer coefficient $k_{i,t}$ [2].

$$\frac{1}{k_{i,t}\rho_m} = \left(\frac{1}{k_{i,m}\rho_m} + \frac{\gamma_i}{k_{i,s}\rho_s K_i f_i}\right) \quad (2)$$

where K_i denotes the equilibrium constant $K_i = \frac{\omega_{i,s}^{\star}\gamma_i}{\omega_{i,m}^{\star} f_i}$ between the two phases for species i. Here γ_i is the raultion activity coefficient for the oxide of species i and f_i is the henrian activity coefficient of species i in the melt. By assuming that the slag and metal is completely mixed the mass balance can be expressed as [2]:

$$-m_m\frac{d\omega_{i,m}}{dt} = k_{i,t}\rho_m A_s(\omega_{i,m} - \omega_{i,m}^{eq}) \quad (3)$$

Here m_p $[kg]$ is the mass of phase p, and A_s $[m^2]$ is the reaction area. In Eq. 3 $\omega_{i,m}^{eq}$ is the hypothetical concentration in the metal at equilibrium with the actual concentration in the slag $\omega_{i,s}$ given by $\omega_{i,m}^{eq}K_i f_i = \gamma_i\omega_{i,s}$. Assuming that $\omega_{i,s}|_{t=0} = 0$ and $\lim\limits_{t\rightarrow\infty}\omega_{i,m} = \omega_{i,m}^{\infty}$ Eq. 3 can be written as:

$$\int_{\omega_{i,m}^{in}}^{\omega_{i,m}} \frac{d\omega_{i,m}}{\omega_{i,m} - \omega_{i,m}^{\infty}} = -\int_{0}^{t} \frac{k_{i,t}\rho_m A_s}{m_m}\left(1 + \frac{\gamma_i m_m}{K_i f_i m_s}\right) dt \quad (4)$$

Equation 4 poses a problem due to most of the parameters on the *RHS* are functions with respect to time [3]. There is also an additional problem in the fact that ladle refining of silicon is done while tapping and therefore cannot be considered a true batch process for its full duration. Kero et al. [3] writes that allowing the *RHS* in Eq. 4 to be constant with respect to time will still give an equation which can be useful in comparing the refining kinetics in this system. This is due to the fact that m_m, m_s, A_s and ρ_m all change in the same way with respect to time for every species in the melt allowing it to serve as a crude approximation, according to Kero et al.

Cussler [7] has gathered multiple expressions for k and from this it can be seen that some common dependencies are $k = f(\rho, U, L, D, \sigma, \mu)$. Here $U\left[\frac{m}{s}\right]$ is the terminal rise velocity, $L[m]$ is some characteristic length, $D\left[\frac{m^2}{s}\right]$ is the diffusion coefficient, $\sigma\left[\frac{J}{m^2}\right]$ is the surface energy density, and $\mu\left[\frac{kg}{ms}\right]$ is the dynamic viscosity. This states that k is a property depending not only on the thermophysical properties of the phases involved, but also on the flow and geometries. If these properties can be considered constant or their behavior similar then $k_{i,t}$ might be considered constant, or similar in behavior as well.

An interesting point when looking at Eq. 4 is that the exponential transient behavior of the *RHS*, after integration, is shared with the much more complex CFD models from Ashrafian et al. [6] and Olsen et al. [8] on the same system. In addition it shows a decent fit with the industrial sample set from Kero et al. [3], as seen in Fig. 2, for many elements, like *Ca*.

The experiments performed in this work differ from the industrial case in that a clean metal is used instead of a slag. This leads to the transport of *Al* and *Ca* from the slag, with a known slag concentration, to the clean metal. However, the mass transfer behavior should not be affected by this as the steps of transport are the same in both cases. Clean metal is used due to it being considerably easier to create a synthetic slag with an accurate composition, than doping pure silicon with specific

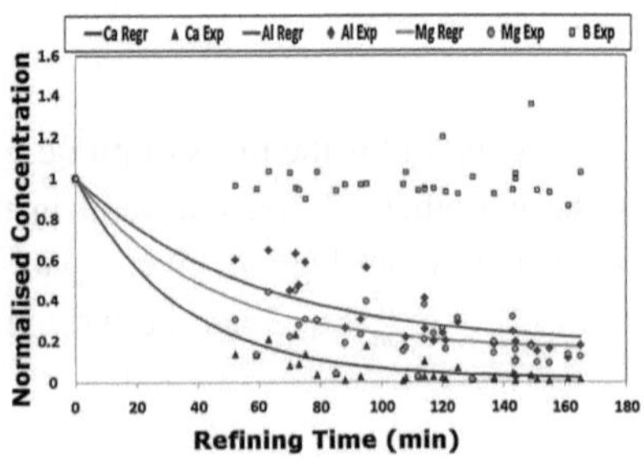

Fig. 2 Normalized *Ca*, *Al*, *Mg* and *B* concentrations as functions of time. From Kero et al. [3]. Printed with permission under the Creative Commons Attribution License as stated in Kero et al. [3]

amounts of *Ca* and *Al*. By applying the same theory as used earlier on this system, with a *RHS* which is not a function of time, gives:

$$\omega_{i,m}(t) = \omega_{i,m}^{\infty}\left[1 - exp\left(-\frac{k_{i,t}\rho_s A_s}{m_m}\left(\frac{K_i f_i}{\gamma_i} + \frac{m_m}{m_s}\right)t\right)\right] \tag{5}$$

Experiment

The apparatus used during the experiment was a graphite tube furnace of an inhouse design, shown in Fig. 3. The furnace was kept under an argon atmosphere at 1.15–1.3 bar to ensure that neither the sample or the graphite was in contact with an oxidizing atmosphere in the hot zone. Two B-type thermocouples were used. The one inserted from the top regulated the temperature, while the second was used as an insurance. This furnace was used due to its special crucible insertion and withdrawal system. It allows the operator to insert and extract samples while the furnace is still running and kept inert. This keeps disturbances to the furnace atmosphere to a minimum, allowing for all the samples in one series to be subjected to as similar conditions as possible.

12 g of silicon pellets were put into 15 IG-110 graphite crucibles with dimensions Ø30 mm/22 × 45 mm and an inner height of 40 mm. The silicon pellets were premelted to minimize inconsistencies in contact area between metal and slag. This gave a smoother surface morphology with a contact area approximately equal to the crucible opening. The furnace was evacuated to ≤ 0.3 mbar and refilled with 5.0 argon. Heating to 1600 °C was started after the internal furnace pressure reached 1.2 bar, at a rate of 40 $\frac{\mathrm{K}}{\mathrm{min}}$. Each crucible was held in the hotzone for 7 min before quenching. The holding time at the chamber separation mechanism was 5 min for the first series, but was later increased to 7 min due to the withdrawal rods heat. 11 g of slag was then added to each crucible, with compositions listed in Table 1. Three different slags were used with five samples in each slag series, totaling 15 samples. Five time steps were chosen and they were 5, 10, 20, 30 and 180 min. The first four time steps measure the mass transfer kinetics, while the 180 min sample was expected to reach equilibrium concentration as proposed by Jakobsson [9]. The same procedure as used above was performed for each sample

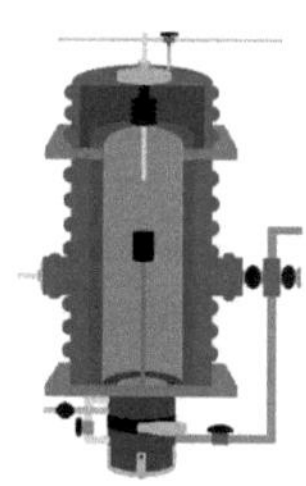

Fig. 3 Graphite tube furnace

Table 1 Intended slag concentrations

Slag ID	SiO_2[wt%]	CaO[wt%]	Al_2O_3[wt%]
CSA255520	55	25	20
CSA254035	40	25	35
CSA404020	40	40	20

set. To keep the experimental conditions as constant as possible each sample set was run on the same day and kept in a desiccator after cooling to avoid contamination.

The silicon used in this experiment had a purity of at least 8 N. Initial master slags were made from commercial oxides with a purity of $\geq$99.5%. Homogeneity was achieved by mixing the oxides in different proportions, melting, quenching and crushing, with the last three steps repeated at least twice. The master slags contained approximately 0.1 wt% Fe_2O_3, with the total initial concentration of all other measured oxides $\geq$0.1 wt%. By mixing the master slags at different proportions the desired slag compositions were obtained. These can be seen in Table 1. To ensure homogeneity the new slag mixtures were heated to 1873 K, kept there for 1 h, before quenched and crushed. The mass loss after heating was only measurable for the 180 min samples where the mass loss was close to 0.6% due to reactions with the crucible.

Before analysis the graphite crucible was removed and the slag and metal separated. Great care was taken to ensure that no metal was present in the slag samples and no slag in the metal samples. As to not contaminate the final product pieces were discarded rather than kept. Over 95 wt% of the metal and at least 60 wt% of the slag was kept after separation. After separation the metal samples were crushed to a fine powder using a tungsten carbide disk mill. The slag samples they were crushed with a polytetrafluoroethylene ball mill.

Both metal and slag were analyzed by ICP-MS to find the concentration of Ca and Al in the metal, and to check for contaminants in the slag. This was done by the HR-ICP-MS lab at the Department of Chemistry at NTNU, with the instrument type Element 2 from Thermo Electronics. Two sub-samples, between 25–45 mg, were taken from each sample and dissolved in a 1.5 mL ultra pure 68% HNO_3 + 0.5 mL 40% HF mixture. After dissolution the samples are diluted, with deionized water, to a final volume between 216–220 mL, giving 0.1 M HNO_3 and 0.23% v/v HF. Three samples of NIST metallurgical silicon standard reference, three standard slag references and three blank samples were analyzed together with the sub samples.

Results

In terms of analytical accuracy Figs. 4, 5 and 6 show the experimental results with a 95% confidence level. There were no other major contaminants found in any of the slag samples beyond what was already noted. The measured values for each sub-sample by ICP-MS are average values from the detector and are delivered with their own relative standard deviation(RSD). Due to this inherent uncertainty the

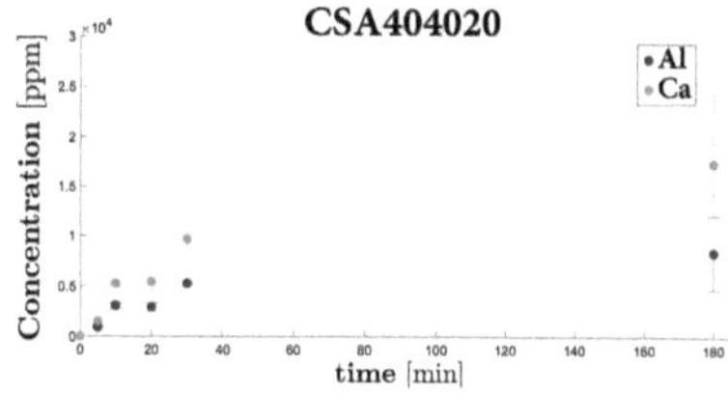

Fig. 4 CSA404020 sample with *errorbars* set at 95% confidence level

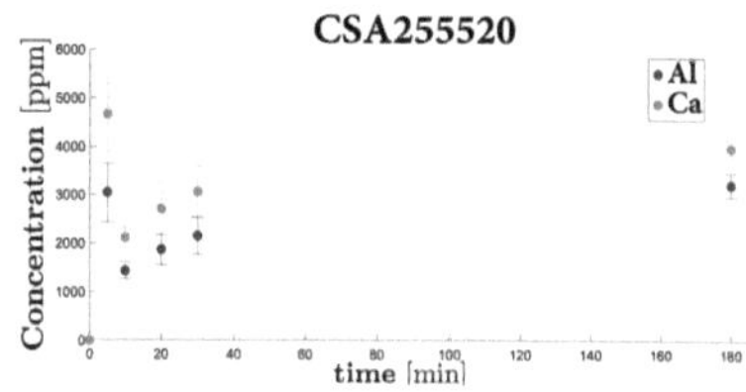

Fig. 5 CSA255520 sample with *errorbars* set at 95% confidence level

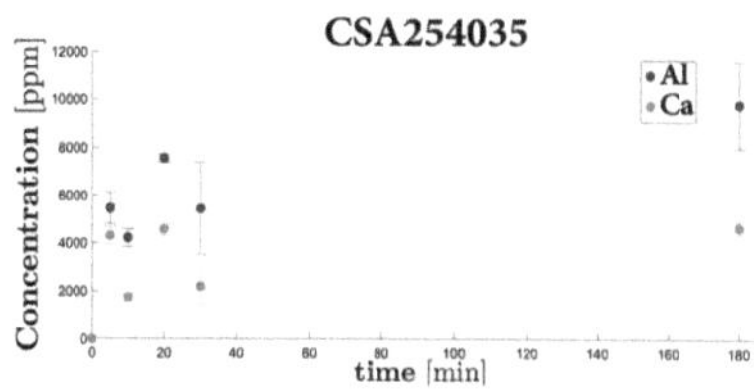

Fig. 6 CSA254035 sample with *errorbars* set at 95% confidence level

STD values are calculated by taking the root of the sum of the variance between sub-samples and the average between the sub-samples own variance. This is expressed in Eq. 6, where the sub-expressions are from the ASTM standard for basic statistics E2586-14.

$$STD = \sqrt{\frac{n\left(\sum_{i=1}^{n} x_i^2\right) - \left(\sum_{i=1}^{n} x_i\right)^2}{n(n-1)} + \frac{\sum_{j=1}^{n}\left(\frac{x_j RSD_j}{100}\right)^2}{n}} \tag{6}$$

Discussion

The experimental method used here has earlier been used by Jakobsson [9], and the method is simple to execute. This method is beneficial due to the fact that there is no flow in the crucible, and the reaction surface should be close to constant. Since the quench is immediate after leaving the hotzone it can be assumed that very little mass transfer happens during cooling giving higher accuracy. One problem with this method is that there is a high loss of slag when separating the two phases, due to the porosity of the slag when machining away the crucible. It is crucial that there are no small pieces of metal in the slag making the loss unavoidable if they are separated by hand. Another experiment using a crucible with an inner diameter of 70 mm and outer height of 150 mm was also performed, but there were considerable difficulty associated with the sampling. Glass tubes with an inner diameter of 4 mm were used inserted trough a graphite tube inserted into the melt. The problem was that considerable amounts of slag entered the sampling tube blocking the metal from entering. This problem is thought to be linked to the high viscosity of the slag and its surface tension. The amount of metal recovered was too small for any reliable analysis. An induction furnace was used for this experiment which induces flows among other things, making it less accurate for predicting specific properties, but it should be useful as another way of generating data to check and validate findings from other experimental methods. It is recommended that another method for sampling is used however.

Samples

If the 180 min entries from Figs. 4, 5 and 6 are compared with equilibrium values calculated by FactSage it can be seen that they are not consistent with the calculated values. Table 2 lists the FactSage and experimental 180 min entries. The Factsage values are calculated using the Fact Oxide- and Fact Light Metals database, at 1600 °C.

From Figs. 4, 5 and 6 it can be seen that while there are some sub-sample pairs which exhibit high degree of variance, the sample setts as a whole there looks to be a high degree of homogeneity between the two sub-samples. This excludes a consistent non-representative sampling of the final product as the culprit for this deviation. The deviation could come from not retrieving enough of the phase in question when separating metal and slag. This might be a reasonable train of thought if the problem lie in the slag, but not for the metal. Another step that might affect the samples could be contamination by the slag when crushing. This does not hold water since the metal and slag were crushed by two different machines.

Operational factors concerning the procedure or the materials used are next in line. The procedure and materials have been tried and tested previously by

Table 2 *Al* and *Ca* concentrations, in [*ppm*], calculated by FactSage, using the Fact Oxide- and Fact Light Metals database, at 1600 °C, in equilibrium with SiO_2-CaO-Al_2O_3 slags, with the experimental values

Sample	CSA404020	CSA255520	CSA254035
Al_{calc}	3900	1000	4500
Al_{exp}	8400	3212	9814
Ca_{calc}	1900	100	280
Ca_{exp}	17339	3970	4660

Jakobsson [9]. Jakobsson did not look at kinetic aspects however so it could be that inserting a cold sample into the hot furnace, and the extra heating associated with this causes some deviations in the early samples. Since the deviation is also found in the later samples this is not considered likely to be the cause of the drastic effect seen here. If this has a large impact then the initial mass transfer should be lower due to a lower temperature, and not much higher as seen from two of the samples. It should be noted that Jakobsson [9] samples show much better agreement for Al then for Ca when compared with the literature.

Next it would be logical to look at the analysis method. ICP-MS is a common method employed when looking for trace elements in metals and slags. The preparation might cause problems due to the fact that *HF* reacts with *Si* and *Ca*. NIST proposes using property data from Chase [10] for CaF_2 and Lyman and Noda [11] for SiF_4. From these it can be seen that SiF_4 is more volatile than CaF_2, so *Si* should react first. Jakobsson [9] found that the *Si* loss was not significant, but if the mixture that was used for all of the samples did cause a loss of *Si* then this could explain the high concentration of *Al* and *Ca*. The calibration mixture for the ICP-MS machine could also be off. When looking at the NIST standard samples the analysis has a low variance for *Al* while the variance of *Ca* is much higher. From this evaluation it seems that calibration problems, or dissolution of the samples might have caused problems when viewing the exact values for the impurities. If this is the case then while the values might be off they should all deviate by approximately the same amount allowing one to view the relative concentration behavior. *Al* does not react strongly with *HF*, compared to *Si* and *Ca*, and the analysis showed consistent values comparing for the NIST standard. Due to this the behavior of *Al* discussed further.

Behavior of Al

If one takes the term assumed constant in Eq. 5, $\frac{k_{i,t}\rho_s A_s}{m_m}\left(\frac{K_i f_i}{\gamma_i} + \frac{m_m}{m_s}\right)$, and views it in context of the experiment performed some assumptions can be made. A_s should be close to constant if one assumes that the quenching is instantaneous, only allowing reactions at 1600 °C. m_m and m_s can also be considered constant as the mass of the crucible did not change to any large degree from before and after the experiment. The concentration of the slag should not change drastically under the experiment allowing for ρ_s and $\frac{K_i f_i}{\gamma_i}$ to also be roughly considered constant. $k_{i,t}$ is then the only parameter left. Sample set CSA404020 shows a behavior akin to the one expected if $k_{i,t}$ is constant. The other samples setts show large deviations from this behavior, and call for a $k_{i,t}$ that changes drastically with respect to time. For this to fit it must allow multiple changes of magnitude with respect to time, to allow for the peaks seen in the experimental data. Since $\lim_{t\to\infty} \omega_i = \omega_i^\infty$, the behavior of $k_{i,t}$ must additionally be of such a form that $\left(\int_0^t k_{i,t} dt\right)\big|_{t\to\infty} \to \infty$. This does not seem likely

Table 3 Slag viscosity calculated from the viscosity model proposed of Kondratiev and Jak [12]

Sample	CSA404020	CSA255520	CSA254035
μ_s [*Pas*]	0.4135	4.0363	1.5086

based on the fact that the phase parameters do not change this drastically to commend this. Fluid flow and geometry changes could have contributed, but there is no significant fluid flow, and the geometries should be constant. Slag viscosity is something that has been looked at and the slag viscosity for the sample setts are shown in Table 3, calculated using the viscosity model proposed by Kondratiev and Jak [12].

Using the same model for small changes in slag concentration shows that the slag viscosity does not change much there either. One thing which is interesting however is that the sample set with the lowest viscosity behaves more akin to the base model, while the sample set with the highest viscosity has the largest initial deviation. This can be seen in Fig. 7, where the concentration of *Al* has been normalized with respect to the 180 min sample. From Fig. 7 it can be seen that if the 5 min sample for the sample set CSA255520 was removed as being an outlier then the behavior of the sample looks somewhat like the batch model [2]. The problem with this still is that the relative initial mass transfer rate from the experimental data is higher with increasing slag viscosity. If all the plot points are considered then it looks like the points oscillate around the curve expected from this model. This relationship is also shared with the industrial data for *Al* from Kero et al. [3]. The industrial data in Fig. 2 show that *Ca*'s behavior closely resembles the batch model, while *Al* has a behavior more akin to the experimental data found here. Since the traditional batch model does not take into consideration the effect the distribution equation between *Ca* and *Al*, reaction 5, this might allow for the behavior seen here. While it is highly speculative it might still be considered.

If one assumes that the metal initially contains no *Al* and *Ca* then the initial mass transfer should be high due to the ease at which the impurity atoms can move across the reaction interface. When the impurity atoms then enter the metal they can quickly diffuse into the bulk away from the reaction surface allowing new atoms to take their place. This leads to reaction (5) becoming unbalanced, which in turn should increase the probability of the offending species atoms to move from the metal into the slag. For this to happen the offending species atoms need to be present at the metal interface in a large enough concentration to restore the balance

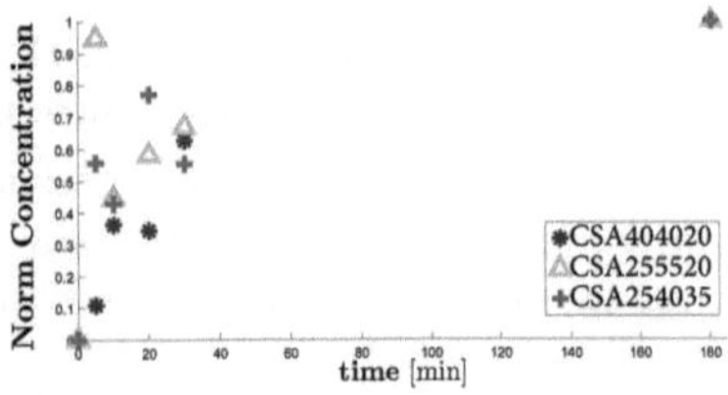

Fig. 7 Normalized concentration of *Al* with respect to the 180 min sample for each sample set respectively

dictated by reaction (5). As the concentration of impurity atoms increases the impurity atoms movements become more constrained increasing the probability that an impurity atom is present at the metal interface. If this is the case then it is not unreasonable that the impurity concentrations might overshoot their expected value. A higher slag viscosity should lower the rate at which the impurity atoms diffuse in the slag, which in turn should increase the residence time of an impurity atom at the slag interface before it can diffuse into the bulk slag. It follows that reaction (5) should then act slower in slags with a higher viscosity allowing a bigger overshoot of impurities in the metal. When the concentration of impurities in the metal becomes high enough it may in turn be favorable to move into the slag causing a new overshoot in the slag. If this occurs continuously where each iteration causes an overshoot which is lower than the previous it can be expressed as an oscillating behavior. For each oscillation there should be a higher readily amount of particles at the interface decreasing the overshoot amount of the species. This thought is somewhat interesting as when one superimposes a damped harmonic motion on to the batch model it provides a better fit for the experimental values, but there is not enough data to support this theory, and with itself it brings other problems.

Conclusions

This work has performed experiments investigating the mass transfer kinetics between *Si* and three synthetic SiO_2-CaO-Al_2O_3 slags, with respect to *Al* and *Ca*, for different time steps at 1600 °C. Different experimental setups were performed, but only one yielded reliable results. The experimental results were found to be differ greatly from the expected behavior and values found in the literature. This was discussed, and it was found that while the values might not be valid the relative behavior of especially *Al* was safe to consider. The behavior was then viewed in light of a traditional batch reactor model [2], and compared to industrial data for oxidative ladle refining from Kero et al. [3]. It was found that *Al* exhibits a behavior which can not be expressed by the batch model alone, and some thoughts on this was shared. To give any definitive answer more experiments need to be performed to be able to say anything for certain.

Further Work

Experiments looking at the reproducibility of the experiments performed in this work are now under way, with further experiments looking at bigger crucibles in the works. For the bigger crucibles more novel sampling techniques are being investigated to allow for a more reliable sampling. Additionally some modeling work is being done to investigate the theoretical behavior of the impurities at a slag/metal interface as individual particles witch respect to their movement and reactions.

References

1. Treatise on Process Metallurgy, vol. 1: Process Fundamentals (Ringgold Inc, 2014).
2. T. Abel Engh, *Principles of Metal Refining*, 2nd ed. (Oxford University Pres, 1992).
3. Ida Kero et al., "Refining Kinetics of Selected Elements in the Industrial Silicon Process", *Metallurgical and Materials Transactions B*, 46 (2015), 1186–1194.
4. Kai Tang, "*Thermodynamic Analysis of Oxidative Ladle Refining of Silicon Melt Including Models for Thermophysical Properties of the Silicon Melt and* SiO_2-Al_2O_3-*CaO Slag*" (Report F8532, SINTEF Materials and Chemistry, 2007).
5. Anders Schei, Johan Kristian Tauset and Halvard Tveit, *Production of High Silicon Alloys*, 1st ed. (Tapir Forlag, 1998).
6. Alireza Ashrafian et al., "*A Reactor Model For Ladle Refining of Silicon Metal*" (6th International Conference on CFD in Oil and Gas, Metallurgical and Process Industries, SINTEF/NTNU, Trondheim, Norway, 10–12 June 2008).
7. E. L. Cussler, *Diffusion: Mass Transfer in Fluid Systems*, 2nd ed. (Cambridge University Press, 1997).
8. Jan Erik Olsen et al., "*CFD Modelling of a Reactive Gas Stirred Three Phase Silicon Reactor*" (6th International Conference on CFD in Oil and Gas, Metallurgical and Process Industries, SINTEF/NTNU, Trondheim, Norway, 10–12 June 2008).
9. Lars Klemet Jakobsson, "*Distribution of Boron Between Silicon and CaO-*SiO_2*, MgO-*SiO_2*, CaO-MgO-*SiO_2 *and CaO-*Al_2O_3*-*SiO_2 *Slags at 1600 °C*" (PhD. Thesis, Norwegian University of Science and Technology, 2013).
10. M. W. Chase Jr, "Tables, NIST-JANAF Thermochemical", *J. Phys. Chem. Ref. Data*", 1998.
11. John L. Lyman and Tetsuji Noda, "*Thermochemical Properties of* Si_2F_6 *and* SiF_4 *in Gas and Condensed Phases*", Journal of Physical and Chemical Reference Data, 30(2001), 165–186.
12. Alex Kondratiev and Evgueni Jak, "A Quasi-chemical Viscosity Model for Fully Liquid Slags in the Al_2O_3-*CaO-FeO-*SiO_2 System", *Metallurgical and Materials Transactions B*, 36 (2005), 623–638.

Part V
Poster Session

The Interface Characteristics of High-Temperature Melt of $CaO\text{-}Al_2O_3\text{-}MgO\text{-}SiO_2$ System

Tian Chen, He Zhi-jun, Zhang Jun-hong, Zhao Xing-tong, Pang Qing-hai and Zhan Wen-long

Abstract The interface characteristics were studied, which contained surface tension and interfacial melting of $CaO\text{-}Al_2O_3\text{-}MgO\text{-}SiO_2$ system in low basicity at high temperature. The $CaO\text{-}MgO\text{-}Al_2O_3\text{-}SiO_2$ slag system interface model was simulated by Butler equation and Girifalco-Good equation. At the same time, in the $CaO\text{-}MgO\text{-}Al_2O_3\text{-}SiO_2$ slag, of which the effect of different basicity on the melting behavior and temperature was simulated and calculated. The experimental result indicated that: the surface tension increases with the increase of $A1_2O_3$ content in the slag system, when the basicity < 2.0. But the effect of the slag system will gradually decrease with the Al_2O_3 increasing. The increase of the content of magnesium can improve the melting temperature of slag system and decrease the melting range.

Keywords $Cao\text{-}Al_2O_3\text{-}MgO\text{-}SiO_2$ system · High-temperature melt · Surface tension · Melting process

Introduction

The slag is the complex compound which formed by the oxide as the main body in the metallurgical process. In addition to the metal melt, it is another important product in the process of metallurgy [1]. There are many chemical reactions which are closely related to physical properties of the slag in the process of smelting and casting. The surface tension and interfacial tension of slag are not only important data to studying the interfacial reaction, but also the most important properties of the slag. The surface tension and interfacial tension of slag have important influence on the interfacial reaction, the removal of inclusions and the corrosion of refractory materials [2–4].

T. Chen · H. Zhi-jun · Z. Jun-hong · Z. Xing-tong · P. Qing-hai (✉) · Z. Wen-long
School of Materials and Metallurgy, University of Science and Technology Liaoning, Anshan 114051, China
e-mail: edikitty@126.com

A. Allanore et al. (eds.), *Materials Processing Fundamentals 2017*,
The Minerals, Metals & Materials Series, DOI 10.1007/978-3-319-51580-9_10

In the ironmaking process, Blast furnace slag (BF slag) can help reducing metal elements and controlling chemical composition of pig iron. In the smelting, the slag adhered to the blast furnace lining (BF lining) so that the slag could make certain protective effect between the molten iron and the BF lining. In addition, the softening-melting zone and the dropping zone were appeared by the formation of blast furnace slag in blast furnace. The interval and temperature of softening-melting and dropping zone can be controlled effectively by controlling composition and properties of slag in order to control the smelting strength and capacity of blast furnace [5, 6].

In this paper, $CaO\text{-}SiO_2\text{-}Al_2O_3\text{-}MgO$ slag system was simulated and calculated so as to studying the melting behavior and the interface characteristics of inquiry under low basicity $CaO\text{-}SiO_2\text{-}Al_2O_3\text{-}MgO$ slag system. According to these, a theoretical basis for the research on the development of low basicity slag can be provided.

Experimental

The Interface Model Calculation of $CaO\text{-}SiO_2\text{-}Al_2O_3\text{-}MgO$ Slag System

It was inevitable to study the surface tension and interfacial tension of slag interface when studying the interface reaction of slag. Butler equation was a surface tension calculation equation, which was based on surface energy and chemical potential. Butler equation has been widely used not only in the calculation of the surface tension between slag and melt but also in the change of internal and surface concentration. At the same time, the Girifalco-Good equation has been widely used in the calculation of interfacial tension. The basic mathematical expressions of both equations were as follows [7–10]:

$$\sigma = \sigma_A + \frac{RT}{S_A}\frac{\sigma_A^S}{\sigma_A^b} = \sigma_B + \frac{RT}{S_B}\frac{\sigma_B^S}{\sigma_B^b} = \ldots = \sigma_i + \frac{RT}{S_i}\frac{\sigma_i^S}{\sigma_i^b} \quad (1)$$

$$\sigma_{m-s} = \sigma_m + \sigma_s + 2 \cdot \phi \cdot (\sigma_m \cdot \sigma_m)^{0.5} \quad (2)$$

The calculation model of this paper was calculated by using Butler and Girifalco-Good equations in order to analyze the variation trend of the surface tension of $CaO\text{-}SiO_2\text{-}Al_2O_3\text{-}MgO$ slag system.

Simulation of Melting Behavior

Effect of Different Basicity on Melting of Slag

The Multiphase Equilibrium of different basicity (CaO/SiO_2) and temperature in CaO-SiO_2-Al_2O_3-MgO slag system was calculated by using the FactSage. Fixing base slag Al_2O_3 = 15%, MgO = 8% and basicity variation range was 1.05–1.25 in experiment. The specific experimental program is showed in the following Table 1.

The Trace Components Influence on Slag Melting

In order to understand the trace components effect on the melting temperature of molten slag, the influence of various trace components on melting temperature were studied and analyzed.

The experimental method of section "Effect of Different Basicity on Melting of Slag" was used to calculate the melting temperature and total melting temperature of the slag by using the FactSage so that the effect of melting temperature under different conditions was obtained. All base slag compositions were showed in Table 2.

Result and Discussion

The Analysis of Interface Model Calculation

Under high temperature, there are many structural units in the CaO-MgO-Al_2O_3-SiO_2 slag system. For example:

Simple ion: Ca +, Mg^{2+}, O^{2-}

Molecular compound: $Al_2O_3SiO_2$, $CaO{\cdot}SiO_2$, $2CaO{\cdot}SiO_2$, $3CaO{\cdot}SiO_2$, $MgO{\cdot}SiO_2$, $2MgO{\cdot}SiO_2$, $MgO{\cdot}Al_2O_3$, $3CaO{\cdot}Al_2O_3$, $12CaO{\cdot}7Al_2O_3$, $CaO{\cdot}Al_2O_3$, $CaO{\cdot}2Al_2O_3$, $CaO{\cdot}6Al_2O_3$, $3Al_2O_3{\cdot}2SiO_2$, $CaO{\cdot}MgO{\cdot}SiO_2$, $CaO{\cdot}MgO{\cdot}2SiO_2$, $2CaO{\cdot}MgO{\cdot}2SiO_2$, $3CaO{\cdot}MgO{\cdot}2SiO_2$, $2CaO{\cdot}Al_2O_3{\cdot}SiO_2$, $CaO{\cdot}Mg{\cdot}2SiO_2$

Let $N_1 = N_{CaO}$, $N_2 = N_{MgO}$, $N_3 = N_{Al2O3}$, $N_4 = N_{SiO2}$, $N_5 = N_{CaO{\cdot}SiO2}$, $N_6 = N_{2CaO{\cdot}SiO2}$, $N_7 = N_{3CaO{\cdot}SiO2}$, $N_8 = N_{MgO{\cdot}SiO2}$, $N_9 = N_{2MgO{\cdot}SiO2}$, $N_{10} = N_{MgO{\cdot}Al2O3}$,

Table 1 Composition of experimental slag

Number	Al_2O_3/%	MgO/%	CaO/SiO_2
1#	15	8	1.05
2#	15	8	1.10
3#	15	8	1.15
4#	15	8	1.20
5#	15	8	1.25

Table 2 Composition of experimental slag

Number	CaO	SiO_2	Al_2O_3	MgO	CaO/SiO_2
1#	42.00	40.00	10.00	8.00	1.05
2#	40.98	39.02	10.00	10.00	1.05
3#	39.95	38.05	10.00	12.00	1.05
4#	39.44	37.56	15.00	8.00	1.05
5#	38.41	36.59	15.00	10.00	1.05
6#	37.39	35.61	15.00	12.00	1.05
7#	36.88	35.12	20.00	8.00	1.05
8#	35.85	34.15	20.00	10.00	1.05
9#	34.83	33.17	20.00	12.00	1.05
10#	43.86	38.14	10.00	8.00	1.15
11#	42.79	37.21	10.00	10.00	1.15
12#	41.72	36.28	10.00	12.00	1.15
13#	41.19	35.81	15.00	8.00	1.15
14#	40.12	34.88	15.00	10.00	1.15
15#	39.05	33.95	15.00	12.00	1.15
16#	38.51	33.49	20.00	8.00	1.15
17#	37.44	32.56	20.00	10.00	1.15
18#	36.37	31.63	20.00	12.00	1.15
19#	45.56	36.44	10.00	8.00	1.25
20#	44.44	35.56	10.00	10.00	1.25
21#	43.33	34.67	10.00	12.00	1.25
22#	42.78	34.22	15.00	8.00	1.25
23#	41.67	33.33	15.00	10.00	1.25
24#	40.56	32.44	15.00	12.00	1.25
25#	40	32	20.00	8.00	1.25
26#	38.89	31.11	20.00	10.00	1.25
27#	37.78	30.22	20.00	12.00	1.25

$N_{11} = N_{3CaO \cdot Al2O3}$, $N_{12} = N_{12CaO \cdot 7Al2O3}$, $N_{13} = N_{CaO \cdot Al2O3}$, $N_{14} = N_{CaO \cdot 2Al2O3}$, $N_{15} = N_{CaO \cdot 6Al2O3}$, $N_{16} = N_{3Al2O3 \cdot 2SiO2}$, $N_{17} = N_{CaO \cdot MgO \cdot SiO2}$, $N_{18} = N_{CaO \cdot MgO \cdot 2SiO2}$, $N_{19} = N_{2CaO \cdot MgO \cdot 2SiO2}$, $N_{20} = N_{3CaO \cdot MgO \cdot 2SiO2}$, $N_{21} = N_{2CaO \cdot Al2O3 \cdot SiO2}$, $N_{22} = N_{CaO \cdot MgO \cdot 2SiO2}$.

Σx_i represents the total mole fraction of the component i; Σx is the total mole fraction of the equilibrium system. N_i represents the concentration of the component i. Many chemical reaction equilibriums were consisted in the slag system (superscript b represents the melt inside), such as:

$$(Ca^{2+} + O^{2+}) + SiO_2 = (CaSiO_3) \tag{3}$$

$$K_1 = \frac{N_5^b}{N_1^b N_4^b}, N_5^b = K_1 N_1^b N_4^b$$

$$\Delta G^{\theta} = -81416 - 10.498\mathrm{T\,J/mol\,ect.}$$

By the total mass conservation:

$$\begin{aligned} &N_1^b + N_2^b + N_3^b + N_4^b + N_5^b + N_6^b + N_7^b + N_8^b + N_9^b + N_{10}^b + N_{11}^b + N_{12}^b \\ &+ N_{13}^b + N_{14}^b + N_{15}^b + N_{16}^b + N_{17}^b + N_{18}^b + N_{19}^b + N_{20}^b + N_{21}^b + N_{22}^b = 1 \end{aligned} \quad (4)$$

$$\begin{aligned} \Sigma x_{CaO} = \Sigma x(&0.5N_1^b + N_5^b + 2N_6^b + 3N_7^b + 3N_{11}^b + N_9^b + N_{10}^b + N_{11}^b + N_{12}^b \\ &+ N_{13}^b + N_{14}^b + N_{15}^b + N_{16}^b + N_{17}^b + N_{18}^b + N_{19}^b + N_{20}^b + N_{21}^b + N_{22}^b = 1 \end{aligned} \quad (5)$$

$$\Sigma x_{MgO} = \Sigma x(0.5N_2^b + 7N_9^b + N_{10}^b + N_{17}^b + N_{18}^b + N_{19}^b + 3N_{20}^b) \quad (6)$$

$$\Sigma x_{Al_2O_3} = \Sigma x(N_3^b + N_{10}^b + N_{11}^b + 7N_{12}^b + N_{13}^b + 2N_{11}^b + 6N_{15}^b + 3N_{16}^b + N_{21}^b + N_{22}^b) \quad (7)$$

$$\begin{aligned} \Sigma x_{SiO_2} = \Sigma x(&N_4^b + N_5^b + N_6^b + N_7^b + N_8^b + N_9^b + N_{21}^b + 2N_{16}^b + N_{12}^b + N_{17}^b + 2N_{18}^b \\ &+ 2N_{19}^b + 2N_{20}^b + N_{21}^b + N_{22}^b) \end{aligned} \quad (8)$$

On the surface of the slag, the same structure units were presented. According to the chemical equilibrium and the total mass conservation, the following equations can be expressed in the melt (Superscript s represents on melt surface):

$$\begin{aligned} &N_1^s + N_2^s + N_3^s + N_4^s + N_5^s + N_6^s + N_7^s + N_8^s + N_9^s + N_{10}^s + N_{11}^s + N_{12}^s + N_{13}^s \\ &+ N_{14}^s + N_{15}^s + N_{16}^s + N_{17}^s + N_{18}^s + N_{19}^s + N_{20}^s + N_{21}^s + N_{22}^s = 1 \end{aligned} \quad (9)$$

Ordering $\sigma_1 = \sigma_{CaO}, \sigma_2 = \sigma_{MgO}, \sigma_3 = \sigma_{Al_2O_3}, \sigma_4 = \sigma_{SiO_2}$ to represent the value of surface tension of CaO,MgO,$A1_2O_3$,SiO_2 respectively. And the Butler equation was applied to the CaO-MgO-Al_2O_3-SiO_2 slag system:

$$\sigma_s = \sigma_1 + \frac{RT}{S_1}\ln\frac{N_1^s}{N_1^b} = \sigma_2 + \frac{RT}{S_2}\ln\frac{N_2^s}{N_2^b} = \sigma_3 + \frac{RT}{S_3}\ln\frac{N_3^s}{N_3^b} = \sigma_4 + \frac{RT}{S_4}\ln\frac{N_4^s}{N_4^b} \quad (10)$$

The calculation model of surface tension of the CaO-MgO-Al_2O_3-SiO_2 slag system was constituted by Type (3) to Type (10). Through working out equations, effect of the concentration and the surface tension of the slag (σ_s) were solved.

Figures 1 and 2 are different slag melt surface tension with the change of the basicity. These figures show that the surface tension with the increase of basicity is also increasing.

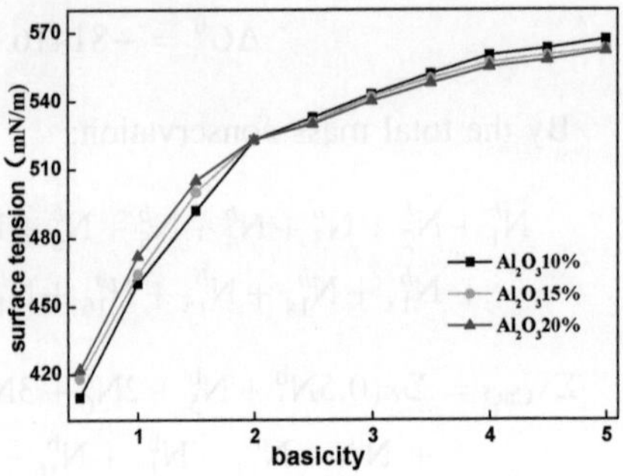

Fig. 1 The surface tension of CaO-MgO-Al_2O_3-SiO_2 slag melt curve with the slag basicity

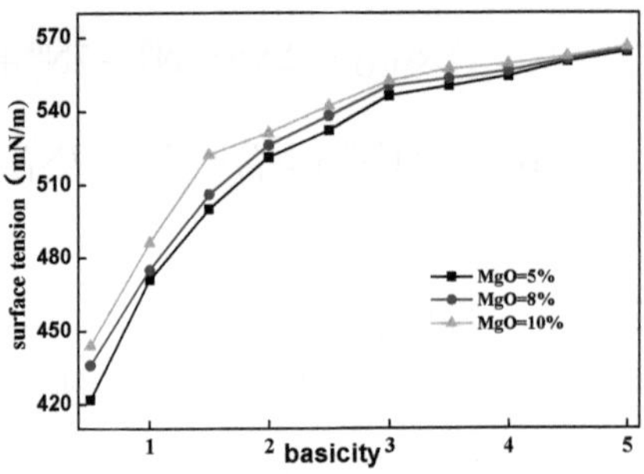

Fig. 2 The surface tension of CaO-MgO-Al_2O_3-SiO_2 slag melt curve with the slag basicity

From the intersection point of the curve in Fig. 1, it can be known that the surface tension increases with the increase of $A1_2O_3$ content in the slag system when the basicity < 2.0. On the contrary, when the basicity > 2.0, the surface tension decreases as the increase of $A1_2O_3$ content in the slag system. Figure 2 shows that the surface tension was improved with the increase of MgO content. However, with the increase of MgO content in high basicity, the change of surface tension gradually slows down.

Analysis of Slag Melting Process

As can be seen in Fig. 3, the law of basicity influence in the melting temperature of the slag is: With the CaO/SiO_2 content increases from 1.05 to 1.25, the beginning of melting temperature and the completely melting temperature of the base slag show a certain degree of increase. When the beginning of the melting temperature improve 10.21 K, the final melting temperature increase 74.5 K. The reason is that it is easy to generate high melting point material such as dicalcium silicate, wollastonite, anorthite which caused the increase of melting temperature with the increase of CaO/SiO_2.

After calculating the trace components effect on the melting temperature of molten slag, data can be seen in Table 3.

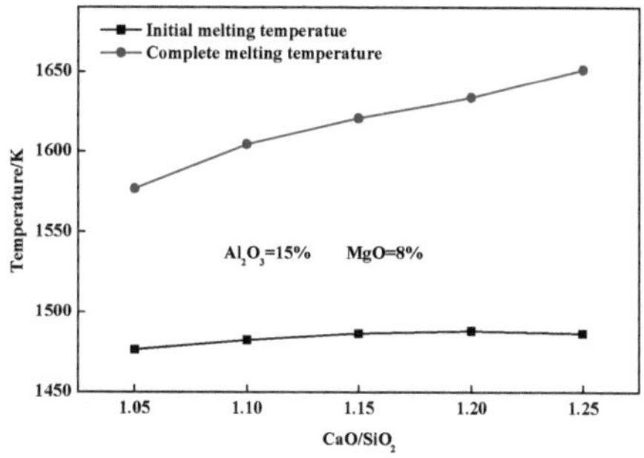

Fig. 3 Effect of CaO/SiO_2 content on melting temperature of slag

Table 3 Melting temperature and complete melting temperature of slag

Number	CaO/SiO_2	Beginning melting/K	Completely melting/K	Maximum melting range/K
1#	1.05	1476.41	1524.49	48.08
2#	1.05	1476.41	1530.77	54.36
3#	1.05	1513.83	1564.09	50.26
4#	1.05	1476.41	1576.82	100.41
5#	1.05	1488.21	1603.32	115.11
6#	1.05	1488.21	1630.65	142.44
7#	1.05	1488.21	1650.39	162.18
8#	1.05	1488.21	1609.58	121.37
9#	1.05	1488.21	1610.15	121.94
10#	1.15	1486.62	1550.24	63.62
11#	1.15	1486.62	1570.66	84.04
12#	1.15	1513.83	1633.28	119.45
13#	1.15	1486.62	1620.81	134.19
14#	1.15	1513.83	1644.33	130.5
15#	1.15	1513.83	1669.22	155.39
16#	1.15	1488.21	1680.84	192.63
17#	1.15	1488.21	1647	158.79
18#	1.15	1513.83	1649.74	135.91
19#	1.25	1486.62	1616.71	130.09
20#	1.25	1513.83	1633.68	119.85
21#	1.25	1513.83	1685.69	171.86
22#	1.25	1486.62	1651.32	164.7
23#	1.25	1513.83	1673.76	159.93
24#	1.25	1513.83	1696.41	182.58
25#	1.25	1513.83	1696.29	182.46
26#	1.25	1513.83	1666.21	152.38
27#	1.25	1513.83	1678.91	165.08

From the experiment, it can be seen that with the increase of CaO/SiO_2, the melting temperature and the melting range of the experimental slag are gradually expanded. The range of difference slag beginning melting temperature was at 37.42 K and completely melting temperature range was at 171.92 K and the minimum melting interval of different calculation samples was 48.08 K and the maximum melting range reached 192.63 K. At the same time, when the content of Al_2O_3 increases and the content of SiO_2 decreases, the melting temperature of slag system will increase, but the effect of the slag system will gradually decrease with the Al_2O_3 increasing. The main reason lies in that the consolidation reaction and the multiphase reaction occurred before the start of melting in the slag system are almost identical. However, in the complete melting, because of the difference of the material component, the CaO/SiO_2 and Al_2O_3 content can increase so as to form a higher melting point, which leads to an increase in the final total melting temperature and a larger melting range. When analyzing the change of magnesium content, the change of the content of magnesium can adjust the melting point and melting zone of the slag system. The increase of the content of magnesium can improve the melting temperature of slag system and decrease the melting range. For the blast furnace smelting process, the size of softening-melting and dropping zone in the blast furnace can be controlled effectively by adjusting the content of MgO, which can achieve the purpose of strengthening smelting.

Conclusion

The surface tension with the increase of basicity is also increasing. With different Al_2O_3 content, the surface tension increases with the increase of $A1_2O_3$ content in the slag system when the basicity < 2.0. On the contrary, when the basicity > 2.0, the surface tension decreases with the increase of $A1_2O_3$ content in the slag system.

The melting temperature and the melting range of the experimental slag are gradually expanded by increase of basicity and content of Al_2O_3 and MgO. However, the effect of the slag system will gradually decrease with the Al_2O_3 increasing. The increase of the content of magnesium can improve the melting temperature of slag system and decrease the melting range.

Acknowledgements This work is financially supported by item sponsored by National Science Foundation of China (NO. 51474124, NO. 51504131, NO. 51504132 NO. 51674139 NO.51604148) and Anshan Science and technology project (NO. 4104).

References

1. Zhang J. Applicability of Mass Action Law to Sulphur Distribution between Slag Melts and Liquid Iron [J]. Journal of University Science and Technology Beijing, 2002, 9(2):90–98.

2. Thermodynamic Calculations on Low Melting Point Area of SiO2-Al2O3-CaO-MgO Inclusion and Its Control. IRON AND STEEL, 2001, 46(1):39–44.
3. H. Ohta, H. Suito. Activities in CaO-MgO-Al_2O_3 slags and deoxidation equilibria of Al, Mg and Ca [J]. ISIJ International, 1996, 36(8): 983–990.
4. WANG Xinhua, CHEN Bin, JIANG Min, et al. Effect of Slag Metal Reaction on Formation of Non metallic Inclusions of Lower Melting Temperature in High Strength Alloyed Structural Steel [J]. IRON AND STEEL, 2008, 43(12):28–32.
5. Xiaodong MA, Geoff WANG, Shengli WU, et al. Phase Equilibria in the CaO–SiO_2–Al_2O_3–MgO System with CaO/SiO_2 Ratio of 1.3 Relevant to Iron Blast Furnace Slags [J]. ISIJ International, 2015, 55(11):2310–2317.
6. LIN Teng-chang, ZHU Rong, WANG Cheng-jie, et al.. Simulation on influence of MgO content on liquid phase in CaO-SiO2-Al2O3-MgO slag [J]. Journal of University of Science and Technology Beijing. 2014, 3 6(1):156–161.
7. YEUM K S, SPEISER R, POIRIER D R. Estimation of the surface tension of binary liquid alloys. Metallurgical Transactions B, 1989, 20 (5):693–703.
8. NAKAJIMA K. Estimation of interfacial tension between phases in the molten iron-slag-inclusion (alumina) system. Journal of the Iron and Steel Institute of Japan, 1994, 80(5): 383–388.
9. YANG Jingjun, WU Yang, CHENG Guoguang. Study on Melting Characteristic of CaO-MgO-Al2O3-SiO2-CaF2 Quinary Slag System [J]. Hot Working Technology, 2014, 43 (7):77–78.
10. YANG Jingjun, CHENG Guoguang. Study on Influence of CaO-Al2O3-MgO Ternary Slag on Fluid Steel Cleanness [J]. INDUSTRIAL HEATING, 2014, 43(1):36–39.

Effect of Modeling Flows on Mixing Time in 40t Ladle with Bottom Gas Blowing Process

Le Wang, Liu Liu and Bingji Yan

Abstract Base on the generalized similarity principle, water model experiments had been conducted to investigate mixing phenomena of argon injection from the ladle bottom with eight types of tuyere arrangement. In this research, the harmonic number (H_0) was considered as an evaluating indicator and four variables were mainly investigated, which were the blowing position, aspect ratio, Froude number and gas flowrate. All of these parameters were dimensionless. Furthermore, A fitting equation was newly proposed for mixing time as a function of four influencing factors. As a result, It was showed that the arrangement of tuyeres has a great effect on mixing time in the ladle and a placement of single tuyere at 2/3R away from the bottom center gave the shortest mixing time.

Keywords Steel ladle · Water modeling · Influencing factor · Mixing time

Introduction

Secondary steel making is today a very important practice in the steel industry because of stringent quality demands regarding steel cleanliness. In the ladle, typically an insert gas like argon is purged through a porous plug or two porous plugs fitted at the bottom to circulate the steel. Gas bubbling in ladles is used to obtain chemical and thermal homogenization, as well as to accelerate the absorption of harmful nonmetallic inclusions into an overlaying slag, etc. [1–4]. Mixing time is a very important parameter in this regard because reducing the mixing time improves the efficiency of the process. It is meaningful to study the mixing phenomenon in ladle refining by water modeling.

L. Wang (✉) · L. Liu
Center of Iron and Steel Research Institute, Beijing, China
e-mail: wangle0077@163.com

B. Yan
Soochow University School of Iron and Steel, Suzhou, China

A. Allanore et al. (eds.), *Materials Processing Fundamentals 2017*,
The Minerals, Metals & Materials Series, DOI 10.1007/978-3-319-51580-9_11

Helle proposed [5] an empirical formula by dimensionless analysis, which showed the relationship between the mixing time and the ladle bottom diameter, the liquid depth, surface tension. According to them, the equation developed in this study for the circulation of the mixing time required during gas rinsing could be used with sufficient accuracy for production purpose.

Some researchers had reported on the various factors which influence mixing time [6]. They were ladle aspect ratio (H/D), presence of upper buoyant phase (slag), nature of motion of the liquid phase, purging location and tracer addition point. Trukoglu and Farouk had demonstrated that both vessel aspect ratio and gas injection play crucial roles in the flow structure in the bath and hence affected mixing time [7].

The purpose of the present research was to study the mixing characteristics in gas injected molten metal baths. The effects of vessel aspect ratio (H/D), the tuyere positions, the flow rate and liquid-gas density ratio on the mixing efficiency were investigated. We carried out water model experiments and proposed an equation for the mixing time.

Experimental

Similarity Considerations

In the gas-stirred system, the fluid flow and mixing of the bathe are caused by momentum transfer between blown gas and liquid. The flow rate of injected gas in the model can be determined by the modified Froude number which was defined by

$$\mathrm{Fr}' = \rho_g u^2/(\rho_l \mathrm{gH}) \tag{1}$$

where H is the height of the ladle.

For N tuyeres with equal inner diameter d, the superficial velocity u was given by

$$\mathrm{u} = 4\mathrm{Q}/(\pi d^2 \mathrm{N}) \tag{2}$$

where Q is the flowrate.

Substitute Eq. (2) into Eq. (1), we obtain

$$\mathrm{Fr}' = 1.621\rho_g Q^2/(\rho_l N^2 d^4 gH) \tag{3}$$

If we assume both of the full scale ladle and water model using the same number tuyeres the following equation can be used to determine the gas flowrate in the model.

$$(Q_1/Q_2) = \left(\frac{\rho_{g2}}{\rho_{g1}}\right) \cdot \left(\frac{\rho_{l1}}{\rho_{l2}}\right) \cdot \left(\frac{H_1}{H_2}\right) \cdot \left(\frac{d_1}{d_2}\right)^4 \quad (4)$$

where the subscripts 1 and 2 stand for model and full scale ladle, respectively.

Experimental Apparatus

Figure 1 showed a schematic of the experimental apparatus. Experimental work was carried out in a one-third scale water model of typical 40-tonne ladle. The main variables involved in the present consideration were listed in Table 1. Based on Miao Yong Zhu and Takeo Inomoto's research [4], we focused the effect of single tuyere or double tuyeres on mixing time combined with the design of the domestic ladle bottom. Eight kinds of bottom blowing arrangements with universal representation were designed in this study. Figure 2 showed different configurations of

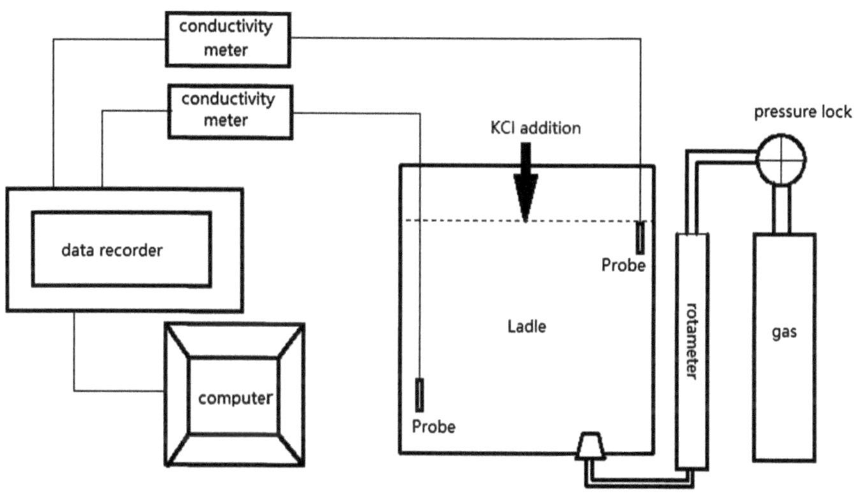

Fig. 1 Experimental apparatus

Table 1 The main variable associated with present studies

	Full scale ladle (40t)	Water model (≈1:3)
Bottom diameter (mm)	2065	596
Top surface diameter (mm)	2289	662
Height (mm)	2482	730
Diameter of tuyere (mm)	80	27
Density of liquid (kg/m^3)	7000	1000

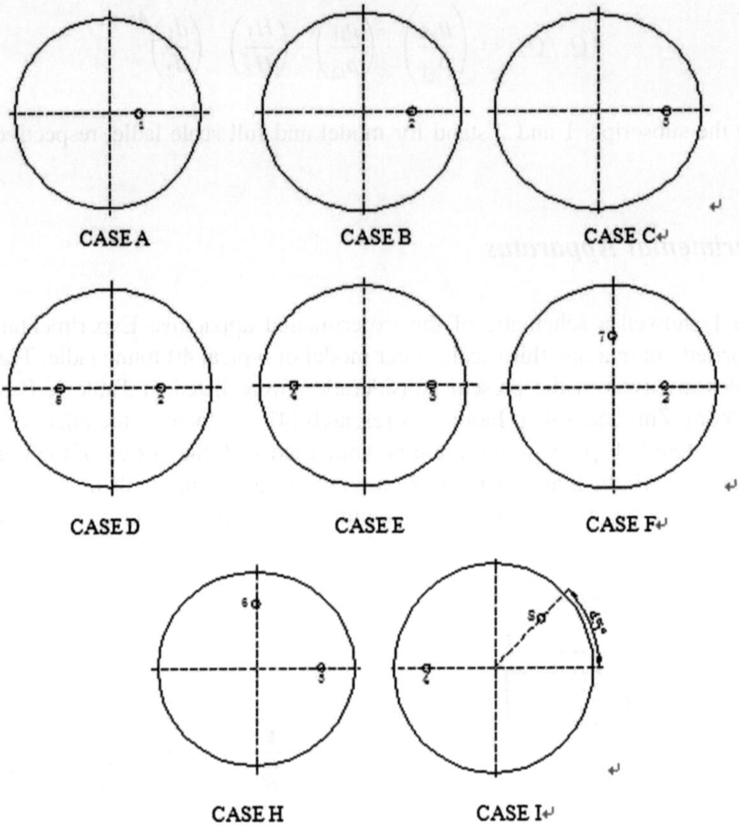

Fig. 2 Eight types of tuyeres arrangement in the ladle

the tuyeres used in experiment. The ladle was filled with tap water with the help of a pump and compressed gas was purged at a definite scaled down rate into the vessel through the bottom tuyeres. The gas flowrate was maintained constant for a particular experiment was with the help of rotameters. In the Fig. 2, tuyere 1 was at the radius of 1/3; tuyeres 2, 5 and 7 were at the half radii in the ladle ($\pm R/2$); tuyeres 3, 4, 6 and 8 were at the radius of 2/3.

In this research, nitrogen and hydrogen were used. The gas purging rate in the actual ladle in the plant ranges between 0.4 and 4 m^3/h which corresponds to 12–120 L/h of nitrogen and hydrogen flowrates for one-third scale water scale water model experiments. KCl solution of 1 N strength was used as the tracer and 50 ml of tracer solution was added for each experiments. An electrical conductivity probe was placed at 200 mm above the bottom and at 3R/4 from the center of the ladle. Another one was placed at 200 mm below the liquid level and also at 3/4R from the center of the ladle. The two probes were fixed on a straight line. The conductivity of the water was detected by the electrical conductivity meter which was connected to a personal computer through a pico log data acquisition system.

Before starting the experiments the probes were calibrated. The mixing time was defined as the time required from the instant of tracer addition until the conductivity meter displayed consistent. Mixing times were measured three times for a given set of experimental conditions, the mean value then being reported as the measured mixing time.

Experimental Result and Discussion

Dimensional Analysis

A large number of studies carried out to data confirms that flow phenomena in gas stirred ladle systems were essentially predominated by the inertial and gravitational forces (e.g. Froude dominated). On the other hand, mixing, from a fundamental stand point, depends on flow and turbulence parameters in the system. Consequently, it is difficult to conceptualize any inter-dependence between mixing times and physical properties of the system (e.g., interfacial tension and viscosity), particularly in the absence of a second buoyant phase [8]. Thus, as a first approximation, assuming flow phenomena to be essentially Froude dominated and furthermore, considering mixing time is influenced by aspect ratio (H/D), the tuyere positions, the flow rate and liquid-gas density ratio, etc. According to Buckingham *pi* theorem [9], it is necessary to determine the number of dimensionless parameters into which the variables may be combined. The relationship between them can be written as follows:

$$\varphi(\tau) = \mathrm{F}\left(\mathrm{D}, \mathrm{W}, \mathrm{dn}, \rho_l, \rho_g, \mathrm{g}, \mathrm{u}, \mathrm{H}\right) \tag{5}$$

Equation (5) is also represented in the following dimensionless form:

$$H_0 = K \cdot (Fr')^a \cdot \left(\pi_\rho\right)^b \cdot (\pi_H)^c \cdot (W)^d \tag{6}$$

where K, a, b, c and d are undetermined coefficient and Fr′, π_ρ, π_H and W are dimensionless parameters.

$$H_0 = u \cdot \tau / D \tag{7}$$

$$\pi_\rho = \rho_l / \rho_g \tag{8}$$

$$\pi_H = H/D \tag{9}$$

Substituted Eqs. (1) and (6)–(8) into (5), Eq. (5) can be rewritten in a dimensionless form as follow:

$$\left(\frac{u\cdot\tau}{D}\right) = K\cdot(Fr')^{a}\cdot\left(\frac{\rho_l}{\rho_g}\right)^{b}\cdot\left(\frac{H}{D}\right)^{c}\cdot(W)^{d} \tag{10}$$

where W is dimensionless value about tuyere configuration. It represents the values of each tuyere position.

The Equation of the Mixing Time

According to Eq. 5, the blowing positions should be converted to dimensionless values. In this study, we used the graphic method to making the blowing position dimensionless [10]. Table 2 showed all of the nondimensionalized blowing position.

In this study, Through stepwise regression analyzing of the data, a relationship between H_0 and H/D, W, Fr′ were found. The equation was described as follows:

$$\lg H_0 = 0.64 + 0.438\lg W + 0.27\lg(Fr' \times 10^4) + 0.06\lg\pi_\rho + 0.29\lg\pi_H \tag{11}$$

Equation (11) was also represented in the following dimensionless form:

$$H_0 = 4.36\cdot W^{0.44}\cdot\left(Fr' \times 10^4\right)^{0.27}\cdot\pi_H^{0.29}\cdot\pi_\rho^{0.06} \tag{12}$$

Equation (12) clearly indicated the index of every influence factor. The size of the index determined the importance of the factor to H_0. It could be seen here that the significant of the influence factor to H_0 were in order of the blowing position, the aspect ratio, the modified Froude number and the liquid-gas density ratio. Among them, the blowing position was the most remarkable.

Table 2 The dimensionless of the blowing positions

Positions	Case A	Case B	Case C	Case D	Case E	Case F	Case H	Case I
$\lg H_0$	2.717	2.665	2.640	2.696	2.682	2.740	2.723	2.820
lgW	0.401	0.196	0.100	0.317	0.263	0.490	0.423	0.800
W	2.517	1.570	1.259	2.076	1.832	3.088	2.651	6.310

The Effect of Configuration of Tuyeres on Mixing

For the sake of comparison, the mean value of measured mixing time was obtained from water model experiment of the different arrangement of tuyeres as shown in Fig. 3. Figure clearly indicates that the shortest mixing time was obtained in the case of single tuyere placed at 2/3R off center (Case C). Under the circumstance of two-tuyere, Case D was found to be the best arrangement that two tuyere were opposite placed at half radii in the ladle. Case H was found to be the worst.

The Effect of Gas Flowrate on Mixing

According to Eqs. 3 and 6, H_0 represented the nondimensionalized mixing time and Fr′ were dimensionless value that have a relationship with gas flowrate. As could be seen from Fig. 4, H_0 as a function of the $Fr'/10^{-4}$ appeared growing trend.

The tendency for predicted mixing time to change with the different gas flowrate was shown in Fig. 5. In general, it could be concluded that the mixing time decrease with increasing gas flowrate. Especially in the case of small gas flowrate, mixing time drastically decreased when the gas flowrate increased. However, the effect was not impressive when the gas flowrate was above 80L/h. This may be due to some reversed flows.

Compared Figs. 4 and 5, when the flowrate increased, H_0 became larger and the mixing time became shorter. The reason for this phenomenon was that the magnitude of the gas flowrate varied much larger than the change of the mixing time.

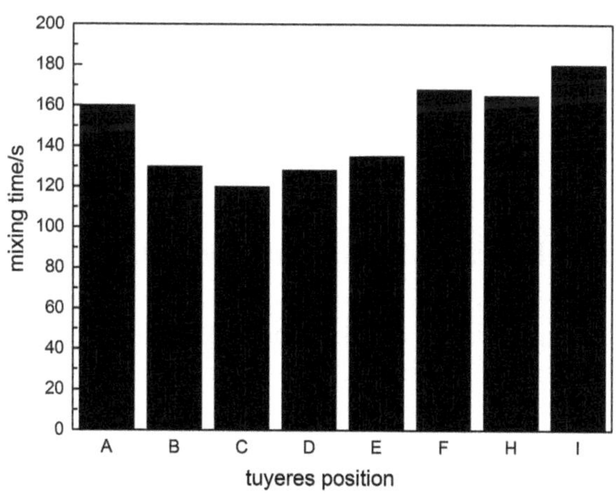

Fig. 3 Relationship between tuyere configuration and the mixing time

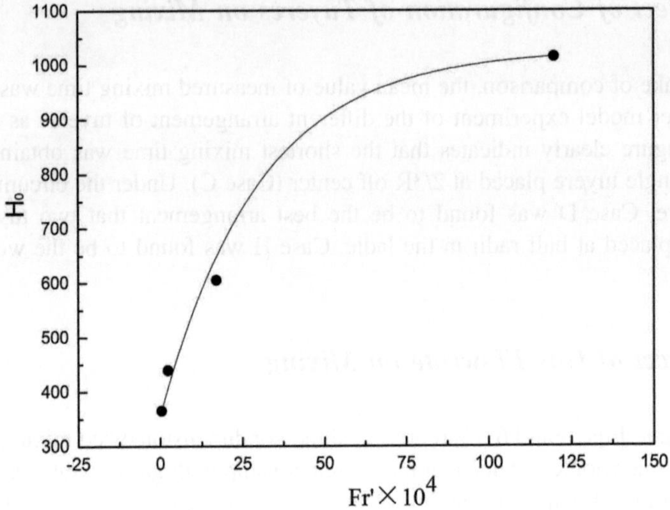

Fig. 4 Relationship between H_0 and Fr′

Fig. 5 Effect of flowrate on mixing time

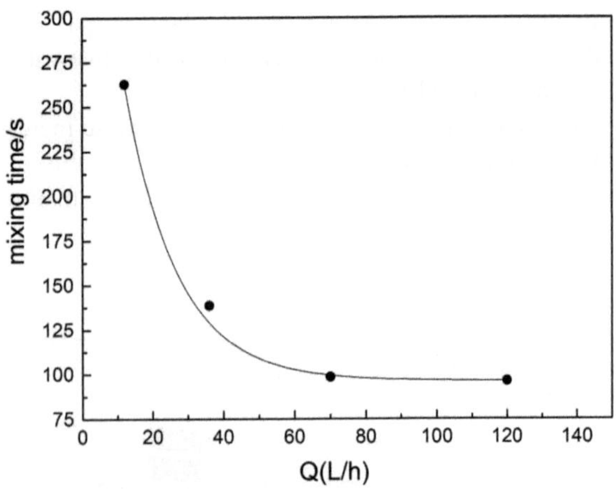

The Effect of Aspect Ratio on Mixing Time

Measured values of mixing time for different aspect ratios(H/D) were shown in Figs. 6 and 7 represented the tendency for H_0 to change with different aspect ratios (H/D). When the aspect ratio increased, the depth of molten pool increased, and the buoyancy of bubbles would be with the greater potential energy. What's more, it benefited for increasing the mixing power of the gas. But, the increasing depth of molten pool led to increasing quality of molten steel, decreasing the unit stirring power and making the mixing became longer.

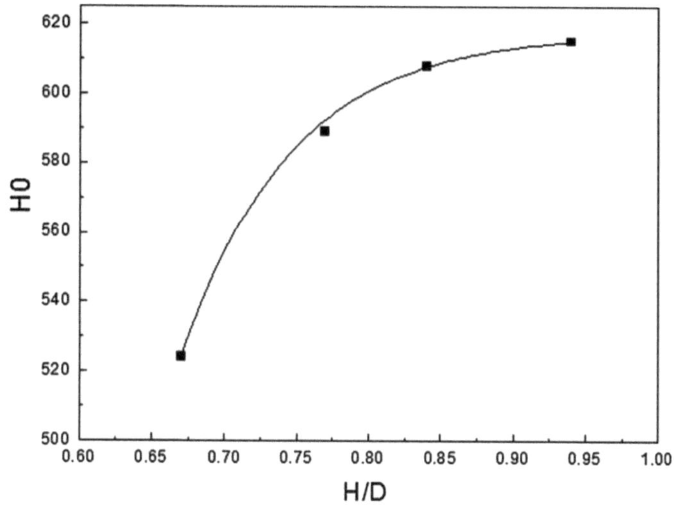

Fig. 6 Relationship between H_0 and aspect ratio

Fig. 7 Effect of aspect ratio on mixing time

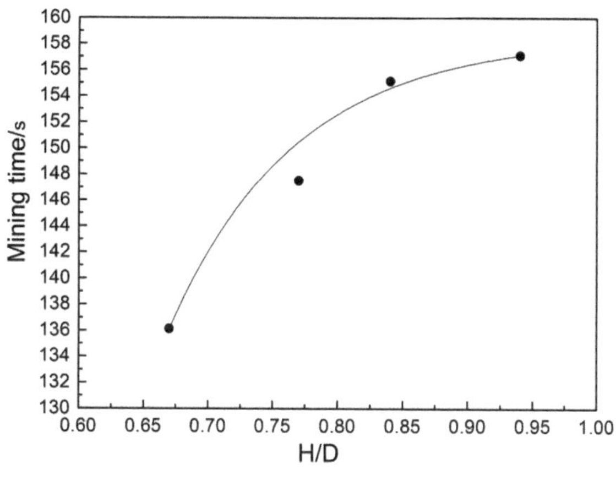

Conclusion

The fluid flow and mixing phenomena in argon-stirred ladles with one or two tuyeres were studied with the help of physical modeling. Considering the effects of tuyere arrangement, injected gas flowrate and aspect ratio on mixing, a proper configuration of tuyeres for the minimum mixing time has been proposed. On the basis of above results and discussions, the following conclusion can be drawn.

(1) The following empirical equation was proposed for mixing characteristic in the ladle. Under certain conditions, the relationship for mixing time could be given by this equation.

$$H_0 = 4.36 \cdot W^{0.44} \cdot \left(Fr' \times 10^4\right)^{0.27} \cdot \pi_H^{0.29} \cdot \pi_\rho^{0.06}$$

(2) In the study, the parameters in the equation were dimensionless. The conclusions were also applicable to the ladle which had a similar size.
(3) Water model experiment showed that the shortest mixing time was obtained in the case of single tuyere placed at 2/3R off center. Under the condition of dual tuyeres, two tuyeres opposite placed at half radii in the ladle was seen as the best arrangement.
(4) Generally, mixing time decreased with increasing gas flowrate, but the effect was not great after 80 L/h. Furthermore, mixing time increased with increasing depth of the pool.

References

1. J. Szekely, *Fluid Flow Phenomena in Metals Processing*, (New York, NY: Academic Press, 1979), 400–408.
2. Koji ABE, Manabu IGUCH, "Mixing Time in Cylindrical Bath Agitated by Plunging Jet," *ISIJ Int.*, 2002, 42(6), 581–587.
3. S. Joo and R. I. L Guthrie, "Modeling flows and mixing in steelmaking ladles designed for single- and dual-plug bubbling operations," *Metallurgical Transactions B,* 23 (1992), 765.
4. Miao Yong Zhu, Takeo Inomoto, Ikuo Sawada, and Tse-chiang Hsiao, "Fluid Flow and Mixing Phenomena in the Ladle Stirred by Argon through Multi-Tuyere," *ISIJ Int.,* 5 (1995), 472–479.
5. Helle, L. W, "The calculation of the time required to mix liquid metal in a ladle by gas rinsing," *J. S. Afr. Inst. Min. Metall.* 81 (1981), 329–337.
6. V. Singh, C. Bhanu, H. J. Billimoria, "Analysis of mixing phenomena in single and double tapered ladles through physical modelling," *TATA STEEL Jamshedpur.*
7. H. Turkoglu, B. Farouk, "Mixing Time and Liquid Circulation Rate in Steelmaking Ladles with Vertical Gas Injection," *ISIJ Int.*, 12 (1991), 1371–1380.
8. Kamalesh Mandal, Dipak MAZUMDAR, "Dimensional Analysis and Mixing Phenomena in Bubble Stirred Ladles," *ISIJ Int.*, 10 (1998), 1150–1152.
9. Buckingham E, "On Physically Similar Systems: Illustrations of the Use of Dimensional Equations," *Physical Review,* 4 (1914), 345–376.
10. Han J J, Shi-Qi LI, and Long W U, "Characteristic number equation about describing mixing effect of steel stirred by argon in the ladle," *Journal of Chongqing University*, 35(1) (2012), 76–80.

Study of a Filter-Press Electrochemical Reactor for the Treatment of Industrial Waste

J.M. Martínez-Vázquez, P.A. Ramírez-Ortega and M. Vargas-Ramírez

Abstract This research reports the effect of hydrodynamics and the distribution of electrical potential in the electrolyte during the electro-recovery of Ag in two prototype of electrochemical reactor filter-press. The input-output smooth (RSM) and the other striated (RST). The conditions of simulation in COMSOL Multiphysics® were laminar flow and mass transfer by convection-diffusion. The concentration of silver deposited on the electrode surface after 2 h of treatment with current values of −35, −65 and −80 mA, was established based on cyclic voltammetry and kinetics using electrodeposition tertiary Nernst- Planck. It was concluded that the reactor geometry influenced the hydrodynamic properties and concentration of Ag recovered. According to the calculations, the velocity profile in the RST was 13.7 times higher with respect to RSM and the recovery of Ag increased a 4% in the RST with respect to RSM to a current value of −80 mA.

Keywords COMSOL® · Electro-recovery · Hydrodynamic · Nernst-Planck · Cyclic voltammetry

Introduction

Modeling and simulation are effective ways to predict the behavior of electrochemical system [1]. Commercially software packages have been developed; to model complex geometries, for example, Elsyca® and COMSOL Multiphysics®.

J.M. Martínez-Vázquez · M. Vargas-Ramírez
Universidad Autónoma del Estado de Hidalgo, Área Académica de Ciencias de la Tierra y Materiales. Carretera Pachuca-Tulancingo km. 4.5. Colonia Carboneras, Mineral de la Reforma, 42184 Pachuca, Hidalgo, Mexico

P.A. Ramírez-Ortega (✉)
Centro de Desarrollo de Nanotecnología, Universidad Tecnológica de Tulancingo, Área de Electromecánica Industrial, Camino a Ahuehuetitla #301 Col. Las Presas, 43642 Tulancingo, Hidalgo, Mexico
e-mail: pedro.ramirez@utec-tgo.edu.mx

A. Allanore et al. (eds.), *Materials Processing Fundamentals 2017*,
The Minerals, Metals & Materials Series, DOI 10.1007/978-3-319-51580-9_12

In this work the fluid transport modules, electroanalysis, tertiary current distribution, transport of diluted species and electrodeposition of software COMSOL 5.0 Multiphysics® were employed.

One of the most important aspects in the design of the electrochemical cells is the distribution of current density in the electrolyte and the electrodes [2]. A non-uniform current density is detrimental, resulting in a nonuniform deposition, high energy consumption [3] and possibly, undesirable side reactions; models of current density distributions are: primary, secondary and tertiary. The Nernst-Planck tertiary interface is chosen to model the electrodeposition because it incorporates transport of species through the diffusion, migration and convection and therefore is able to describe the effects of composition throughout the electrodeposition process well applies to the equations of electroneutrality what describing all the species [4]. This involve that all charged species in the electrolyte specified in the simulations, except those species that are present in very low concentrations and therefore do not contribute to the current balance. Electrodeposition is the process by which metallic ions present in the electrolyte are reduced to metal which is deposited on an electrode, to which is supplied a current or potential electrical. A basic electrochemical cell consists of an electrolyte with two immersed electrodes to which a load is applied direct current. In the negatively charged electrode, or cathode, positively charged ions reduced to form a solid (Ag^0) or evolve as gas (H_2). While, at the anode, negatively charged, the ions are oxidized. The main reactions occurring during the electrorecovery of silver (Ag) dissolved in nitric acid (HNO_3), can be written as [5]:

At the anode:

$$Ag^+ \rightarrow Ag^{2+} + e^- \tag{1}$$

$$Ag^{2+} + NO_3^- \rightarrow Ag(II)NO_3^- \tag{2}$$

$$Ag^+ + NO_3^- \rightarrow Ag(II)NO_3^- + e^- \tag{3}$$

At the cathode:

$$HNO_3 + 2H^+ + 2e^- \rightarrow HNO_2 + H_2O \tag{4}$$

$$3HNO_2 \rightarrow HNO_3 + H_2O + 2NO \tag{5}$$

$$HNO_3 + 3H^+ + 3e^- \rightarrow NO + 2H_2O \tag{6}$$

This paper presents the numerical modeling of the effect of hydrodynamics [6] on the momentum transport, cyclic voltammetry, the potential distribution, tertiary current and electrodeposition, on an electrochemical reactor prototype, filter-press [7] used to study the recovery process silver [8]. The conditions of the simulation are laminar flow [9], the mass transfer model was a combination of convection-diffusion equations, and the species concentration dependent of the kinetics.

Mathematical Models [10]

In the first stage of the study, the models were drawn at Parasolid® software, later imported to COMSOL® to determine the hydrodynamic properties of the fluid, selecting a laminar flow. In the next step the electrolyte distribution in the reactor and its effect on the current distribution was studied. Therefore, once the geometry of the reactor established, a study on the distribution of electrolyte in the reactor was performed to ensure current distribution as uniform as possible. Subsequently, was used the module tertiary electrodeposition Nernst-Planck in order to know the amount of silver that was deposited on the steel cathode 304, by simulating the reactor operation for 2 h at room temperature [11]. In Fig. 1, are shown the geometries of the two types of electrochemical reactors simulated, smooth (RMS) and striated (RST).

Hydrodynamic Properties

The momentum transport of fluid flow is described by the Navier-Stokes (Eq. 7) at steady state:

$$\rho(\boldsymbol{u}\cdot\nabla)\boldsymbol{u} = \nabla\cdot\begin{bmatrix} -pI+\eta\left(\nabla\boldsymbol{u}+(\nabla\boldsymbol{u})^{T}\right)- \\ \left(\frac{2\eta}{3}-\kappa_{dv}\right)(\nabla\cdot\boldsymbol{u})I \end{bmatrix} + \boldsymbol{F}\nabla\cdot(\rho\boldsymbol{u}) = 0 \tag{7}$$

where, η is the dynamic viscosity (N s m^{-2}), **u** is the velocity gradient, p is the pressure, ρ the fluid density (kg m^{-3}) and F is the volumetric force (N m^{-3}). At the entrance one must specify a velocity vector (**u**) normal to the contour (**n**):

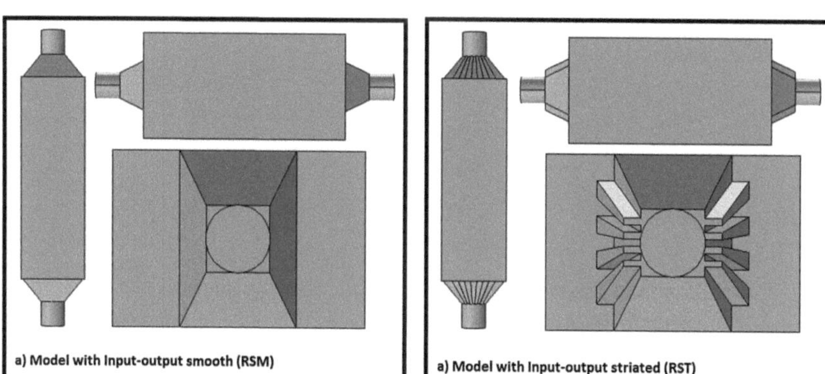

Fig. 1 Models of reactor filter-press used in this research

$$\boldsymbol{u} \cdot \boldsymbol{n} = u_0 \tag{8}$$

In the output contour, the pressure (p) is equal to the initial pressure (p_0) according to the Eq. (3):

$$p = p_0 \tag{9}$$

Finally, in the reactor walls, applies a boundary condition of no slip:

$$\boldsymbol{u} = 0 \tag{10}$$

The mass transport in the reactor is given by equations convection and diffusion as shown in Eq. (11):

$$\frac{\partial c_i}{\partial t} + \nabla \cdot (-D_i \nabla c_i + c_i \boldsymbol{u}) = R_i \tag{11}$$

where, D_i is the diffusion coefficient ($m^2\ s^{-1}$), and R_i denotes the end rate of reaction ($molm^{-3}s^{-1}$). For the boundary conditions, must be specified the input concentrations (c_i):

$$c_i = c_{i,0} \tag{12}$$

In the output, the mass flow through the contour, is dominated by the diffusion process. This assumes that any mass flow which crosses the contour due to diffusion is zero.

$$n \cdot (-D_i \nabla c_i) = 0 \tag{13}$$

Considering the flow of species i (mol $m^2\ s^{-1}$), Eq. (8) yields:

$$\boldsymbol{N}_i \cdot \boldsymbol{n} = c_i \boldsymbol{u} \cdot \boldsymbol{n} \tag{14}$$

Finally, at the reactor walls no mass transport across the contours:

$$\boldsymbol{N}_i \cdot \boldsymbol{n} = 0 \tag{15}$$

Cyclic Voltammetry

The cyclic voltammetry is an analytical technique used to analyze electrochemical systems. The method involves a linear sweep over time of the potential difference between a working electrode and a reference potential from a potential start vertex, and vice versa. The resulting current in the working electrode is plotted against the potential in a voltammogram. The information, on both the electrochemical reactivity and transport properties of an electrochemical system can be obtained simultaneously. The interface of electroanalysis, uses equations of chemical transport for species reactant and product. The domain equation is the diffusion equation (also known as Fick's second law) to describe the transport of substances chemicals of the species electroactive A and B:

$$\frac{\partial c_i}{\partial t} = \nabla \cdot (D_i \nabla c_i) \tag{16}$$

At the upper limit (x = L), a uniform concentration, is assumed equal to the highest concentration of the reactant. The product has a concentration of zero here; on the edge of the electrode (x = 0), the reactive species A, is oxidized (loses an electron) to form product B. By convention, the electrochemical reactions are written in the direction of the reduction:

$$B + e^- \leftrightarrow A \tag{17}$$

The stoichiometric coefficient is −1 for B (the reagent in the direction of the reduction), and +1 for A (the product in the direction of the reducing); the number of electrons transferred, n equals one. The current density for this reaction is given by the electroanalytical Butler-Volmer equation for oxidation:

$$i_{loc} = nFk_0\left(c_A \exp\left(\frac{(n-\alpha_c)F\eta}{RT}\right) - c_B \exp\left(\frac{-\alpha_c F\eta}{RT}\right)\right) \tag{18}$$

where k_0 is the rate constant of the reaction heterogeneous, α_c is the coefficient of transfer and η is the cathodic overpotential on the working electrode. The overpotential is the difference between the potential applied and the equilibrium potential (formal potential of the reduction) of redox couple of species A and B. According to the Faraday laws of electrolysis, the flow of reactive and products are proportional to the density of current expended:

$$-\boldsymbol{n} \cdot \boldsymbol{N}_i = \frac{v_i i_{loc}}{n\boldsymbol{F}} \tag{19}$$

This is expressed in the boundary condition of the electrode surface. The triangular waveform applied to the study of cyclic voltammetry specified in the boundary condition of the electrode surface according to the two vertices of potential, forming a window in potential between −0.4 and 0.4 V, on each side of the potential of reduction on the equilibrium, and a scan rate voltammetric, v (v s^{-1}), which is the rate at which the applied potential changed. The total current is related to the current density by simply multiplying by the electrode area, A:

$$I_{el} = i_{loc}A \tag{20}$$

Current Tertiary Distribution

In the density distribution of tertiary current, mass transport by diffusion, convection and migration are considered. The net charge ion transport is the electroneutrality and the electrolyte supporting with concentration negligible gradients, which means that the potential distribution in the electrolyte can be described by Ohm's Law to introduce a dependence on mass transport in this model, the species oxidized at the anode is mass transport limited and its concentration is, c (mol/m^3) therefore, affect the kinetics of the electrodes. The anodic branch of the Butler-Volmer expression (Eq. 15) at the anode has a concentration dependence.

$$I = A \cdot i_0 \cdot \left\{ exp\left[\frac{(1-\alpha)nF}{RT}(E - E_{eq})\right] - exp\left[-\frac{\alpha nF}{RT}(E - E_{eq})\right]\right\} \tag{21}$$

where, I is the electrode current, i_0 is the density of exchange current, n is the number of electrons involved in the electrode reaction, A is the active surface area of the electrode, T is temperature, R is the constant of gases (J K^{-1} mol^{-1}), α is the coefficient of charge transfer, E and E_{eq} are the cell potential and the equilibrium respectively. The Butler-Volmer equation is based on pure metals electrodes. The expression, now, for the distribution of tertiary current is as shown in Eq. (22).

$$i_a = i_0\left(\frac{c}{c_0}\exp\left(\frac{\eta(1-\beta)F}{RT}\right)\right) - \exp\left(\frac{\eta\beta F}{RT}\right) \tag{22}$$

Here c_0 (mol m^{-3}) denotes a reference concentration and is equal to the inlet concentration and β is the symmetry factor. The above equation is applied to the anode, for the cathode is maintained the model of density distribution secondary current.

$$i_{\mathrm{loc}} = i_0\left(\exp\left(\frac{\eta(1-\beta)F}{RT}\right)\right) - \exp\left(\frac{\eta\beta F}{RT}\right) \tag{23}$$

The overpotential, η, is the difference between the electrode potential ($E_{electrode}$) and the equilibrium potential (E_{eq}) for the reaction in electrode, and is defined as follows:

$$\eta = E_{electrode} - E_{eq} \tag{24}$$

This results in the following expressions for overvoltage at the cathode (η_c) and the anode (η_a) with respect to the potential of the electrochemical cell (E_{cell}), the ionic potential of the liquid in the cathode ($\varphi_{l,\ c}$), in the anode ($\varphi_{l,\ a}$) and the equilibrium potential at the anode ($E_{eq,\ a}$) and cathode ($E_{eq,\ c}$):

$$\eta_c = -\varphi_{l,c} - E_{\mathrm{eq},c} \tag{25}$$

$$\eta_a = E_{\mathrm{cell}} - \varphi_{l,a} - E_{\mathrm{eq},a} \tag{26}$$

In addition, a balance on momentum is introduced to describe convection. In this case, the flow is assumed laminar incompressible stationary, using the Navier-Stokes equations (Eq. 7), the equation for mass transport of the reacting species is:

$$\nabla \cdot (-D\nabla c - zmFc\nabla\varphi + c\mathbf{u}) = 0 \tag{27}$$

where c is the concentration (mol m^{-3}), z the valence, D the diffusivity ($m^2\ s^{-1}$), m the mobility (mol m^2 (s VA)), F is the Faraday constant (A s mol^{-1}), φ the ionic potential and **u** the velocity vector (m s^{-1}). For all boundary limits, conditions non-flow except for the input, output and anode are applied. At the entrance is fixed a concentration. Faraday's law is used to specify the net molar flow at the anode where the species is consumed:

$$N_a = \frac{i_a}{F} \tag{28}$$

Electrochemical Silver Recovery

The electrodeposition simulates the evolution of the accumulation of the silver deposits with time on the electrode surface. The ion transport in the electrolyte occurs by convection and diffusion. The electrode kinetics is described by the expression of Butler-Volmer, dependent concentration [4]. The reaction at the electrode will move the limit in the normal direction with a speed v_{dep} (m s^{-1}) according to:

$$v_{\mathrm{dep}} = \frac{M_{\mathrm{Ag}}}{\rho_{\mathrm{Ag}}} \frac{i_{\mathrm{ct}}}{2F} \tag{29}$$

where M_{AG} is the molecular weight, and ρ_{Ag} the density of silver, respectively. The problem is time-dependent to simulate the electrode deformation for a time t in seconds.

Results and Discussion

Figure 2 shows the velocity profiles of reactor filter-press without applying a potential for a) RSM and b) RST respectively. In both models, the high velocity zones are located in the inlet and outlet of the reactor, the velocity values ($V_{RSM} = 0.309$ m s^{-1} and $V_{RST} = 0.307$ m s^{-1}) and pressure ($p_{RSM} = p_{RST} = 1.61$ kPa) were similar in both reactors. However, when a change in the cell potential of 1.8 V is applied, the velocity profile changes for both models, as shown in Fig. 2c and d. The RSM exhibits a decrease in velocity (VRSM = 0.0254 m s^{-1}) and the velocity profile is similar to showed without application of the potential. The RST exhibits an increase in speed ($V_{RST} = 0.342$ m s^{-1}) in the order of 13.7 with respect to the RSM in the same conditions.

For reducing potentials, the oxidation reaction is not taking place and the generated current is negligible, as shown in Fig. 3. When the potential, move toward the reduction potential of the redox couple, the reaction of oxidation is accelerated and the current increases. Once the oxidation reaction has consumed the reagent on the electrode surface, the current is limited by the rate of transport of Ag^+ to the working electrode. Therefore, a current peak is observed, and at higher potentials, the voltammetric current drops to a potential independent of the speed; this region is called "diffusion controlled" or "controlled transport". In the sweep back toward more reducing potential, conversion of the product, Ag^0, in the original reagent (Ag^+) gave a negative current value (cathode, reductive). The depletion of the reacting species causes a negative peak current and conversion takes place, then, controlled by the diffusion rate. In Fig. 4, the concentration profile of RST reactor, shows a uniform distribution of silver on electrode; whereas in the RSM, the concentration profile focuses on the inputs and outputs of the reactor. The uniform distribution in the RST reactor could likely lead is a purger recovery of silver with respect to the RSM reactor, as seen below.

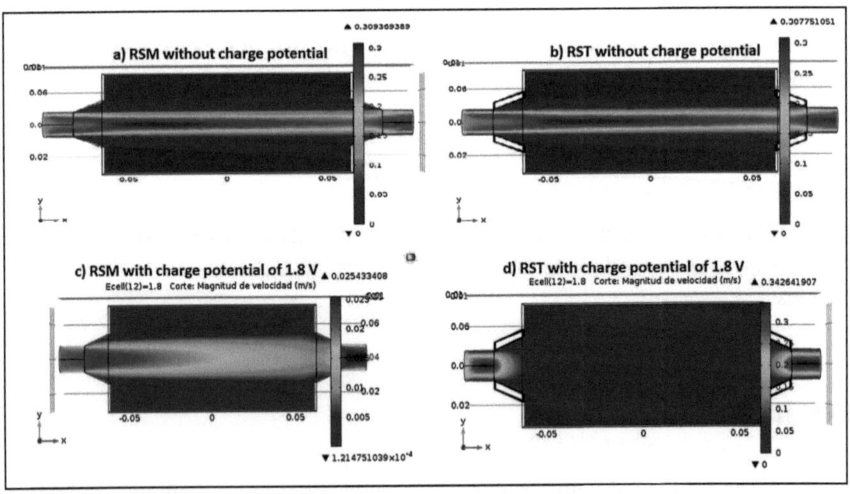

Fig. 2 Velocity profiles without applying a potential **a** RSM and **b** RST. Velocity profile of the reactor when a potential of cell of 1.8 V was applied

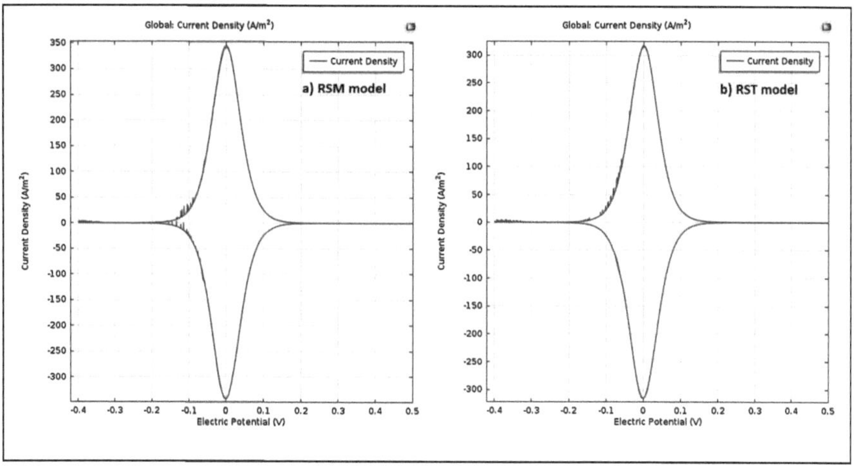

Fig. 3 Cyclic voltammetry of reactor filter-press when to was applied a potential −0.4 to 1.8 V in **a** RSM **b** RST

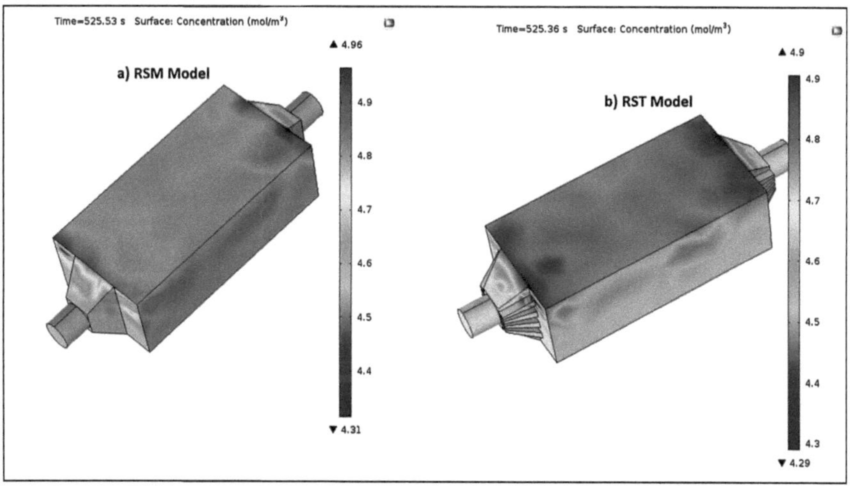

Fig. 4 Concentration profile in the reactor **a** RSM and **b** RST, after applied the study of cyclic voltammetry

In Fig. 5, the effect of reactor model and current distribution in the potential of the electrolyte for the time interval from 0 to 7200 s, indicates that the geometry and current does not significantly affect the distribution of potential in the electrolyte.

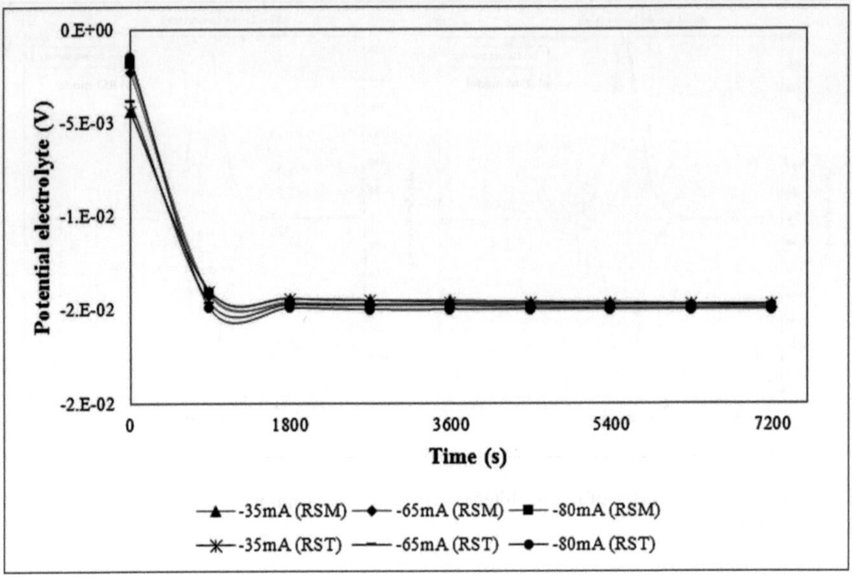

Fig. 5 Profile potential distribution in the electrolyte in reactors RSM and RST for the time interval from 0 to 7200 s to the values of current of −35, −60 and −80 mA

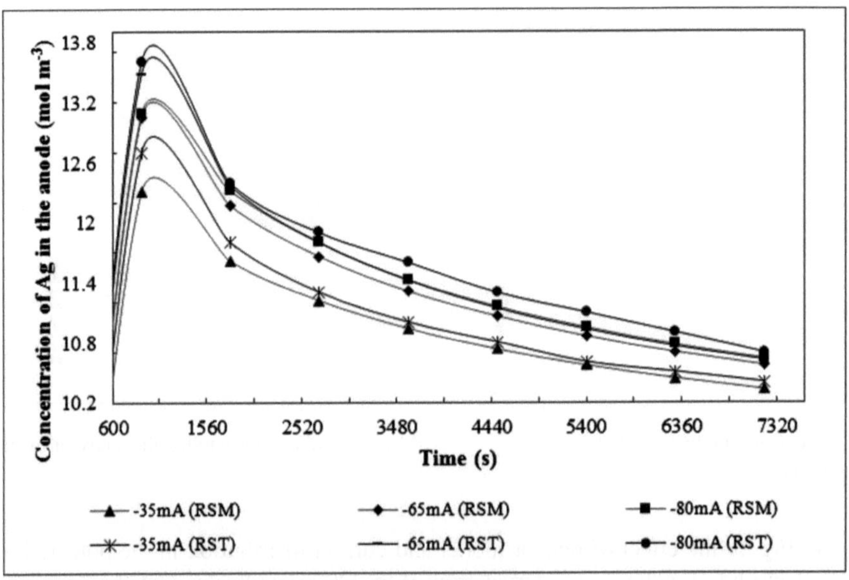

Fig. 6 Concentration profile of the silver electro-recovered for the work of current of −35, −65 and −80 mA in the reactors RSM and RST

The concentration profiles in the RSM and RST reactors, shown a behavior similar in relation to Fig. 6, showing a higher concentration of Ag at a current of −80 mA in the RST reactor at an approximate time of 1080 s.

Conclusions

The geometry of the reactor and electrochemical kinetics are mainly influenced by the hydrodynamic properties of the reactor filter-press and the concentration of the electro silver recovered. The variation in the distribution of the potential in the electrolyte for each current value is 1–2% maximum. Therefore, it can be inferred that the geometry of the reactor contributed with an increase of 4% in of Ag concentration electro-recovered in the RST reactor with respect to the RSM.

Acknowledgements The authors externalize their gratitude to PRODEP, the Universidad Tecnológica de Tulancingo and the Universidad Autónoma del Estado de Hidalgo for the support granted for the development of this research.

References

1. E. P. Rivero, F. F. Rivera, M. R. Cruz-Díaz, E. Mayen, I. González, "Numerical simulation of mass transport in a filter press type electrochemical reactor FM01-LC: Comparison of predicted and experimental mass transfer coefficient," *Chemical Engineering Research and Design*, 90 (2012)., 1969–1978.
2. I. Perissi, C. Borri, S. Caporali, A. Lavacchi, Excerpt the Proceedings of the COMSOL Conference, Milan, Italia, (2009).
3. L. Yu-Ming, L. Shuo-Jem, L. Chi-Yuan, C. Dar-Yuan, "The multiphysics analysis of the metallic bipolar plate by the electrochemical micro-machining fabrication process," *Journal of Power Sources*, 193 (2009), 227–232.
4. E. P. Rivero, M. R. Cruz-Díaz, F. J. Almazán-Ruiz, I. González, "Modeling the effect of non-ideal flow pattern on tertiary current distribution in a filter-press-type electrochemical reactor for copper recovery," *Chemical Engineering Research and Design*, 100 (2015), 422–433.
5. K. C. Pillai, S. J. Chung, I. S. Moon, "Studies on electrochemical recovery of silver from simulated waste water from Ag(II)/Ag(I) based mediated electrochemical oxidation process," *Chemosphore*, 73 (2008), 1505–1511.
6. U. M. López-García, P. E. Hidalgo, J. C. Olvera, F. Castañeda, H. Ruíz, G. Orozco, "The hydrodynamic behavior of a parallel-plate electrochemical reactor," *Fuel*, 110 (2013), 162–170.
7. P. A. Ramírez Ortega, V. E. Reyes Cruz, M. A. Veloz Rodríguez, "Recuperación de plata de desechos radiológicos en un reactor electroquímico tipo prensa sobre un electrodo de acero inoxidable A304 variando la velocidad de flujo" (*Memorias en Extenso del XXIV Congreso de la Sociedad Mexicana de Electroquímica*, Puerto Vallarta, México, 2008).
8. V. Reyes-Cruz, I. González, M. T. Oropeza, "Electro-recovery of gold and silver from a cyanide leaching solution using a three-dimensional reactor," *Electrochemica Acta*, 49 (2004), 4417–4423.

9. A. Frías-Ferrer, I. Tudela, O. Louisnard, V. Sáez, M. Deseada Esclapez, M. I. Díez-García, P. Bonete, J. González-García, "Optimized design of an electrochemical filter-press reactor using CFD methods," *Chemical Engineering Journal*, 169 (2011), 270–281.
10. John Newman and Karen E. Thomas-Alyea, *Electrochemical Systems*, (Hoboken, New Jersey: John Wiley & Sons, Inc., 2004), 7.
11. J. M. Martínez Vázquez, P. A. Ramírez Ortega, M. Vargas Ramírez, "Simulación del proceso de electro recuperación en un reactor tipo prensa mediante COMSOL®" (Paper presented at the XXXI Congreso Nacional de la Sociedad Mexicana de Electroquímica, Monterrey, Nuevo León, 30 May 2016), 336.

Genetic Influence of Mold Corner Structure on the Strand Corner Temperature in Secondary Cooling Zone During Slab Continuous Casting

Sheng Yu, Dengfu Chen, Pei Xu, Mujun Long, Kui Lv and Huamei Duan

Abstract Mold corner structure will influence the strand corner temperature by changing fluid flow and heat transfer of liquid steel at mold corner. A three-dimensional model coupling fluid flow and heat transfer in the mold and a two-dimensional moving-slice heat transfer model in secondary cooling zone was established to investigate slab corner temperature under different mold corner structure: right-angle, big-chamfer, multi-chamfer and fillet. Results show that strand corner temperature increases obviously at mold exit through using chamfered mold. Compared with the corner temperature of right-angle strand, corner temperature of big-chamfered, multi-chamfered and fillet strand are promoted during the straightening process, which will avoid the low ductility temperature zone. When mold corner structure changed from right-angle to big-chamfer, the rise of corner temperature is much higher than that changed from big-chamfer to multi-chamfer or fillet.

Keywords Continuous casting · Mold corner structure · Corner temperature · Secondary cooling · Chamfered mold

Introduction

The corner crack of micro-alloy steel has been a serious problem in slab continuous casting. During the secondary cooling process, especially in straightening segment, the slab may have corner crack under the influence of the complex thermal stress and straightening stress. It will seriously affect the production and raise the cost. By increasing or reducing the heat transfer [1] around the strand corner to avoid the low

S. Yu (✉) · D. Chen · P. Xu · M. Long · K. Lv · H. Duan
College of Materials Science and Engineering, Chongqing University,
Chongqing 400044, China
e-mail: jscquys@163.com

D. Chen
e-mail: chendfu@cqu.edu.cn

A. Allanore et al. (eds.), *Materials Processing Fundamentals 2017*,
The Minerals, Metals & Materials Series, DOI 10.1007/978-3-319-51580-9_13

ductility temperature zone [2], the corner crack has been reduced but is hard to be avoided.

Chamfered mold is an effective way to improve the strand corner temperature through changing the mold corner structure. In 1994, British Steel [3] used a special mold with bevel angle to conduct industrial experiments. Then some other steel-works developed slab chamfered mold, but was not very successful [4]. In recent years, some organizations in China, like Central Iron and Steel Research Institute, have conducted a series of industrial experiments [5] and mathematical simulation [6] to promote the chamfered mold. However, these work are focused on the big-chamfered mold. The influence of different mold corner structure on the corner temperature in secondary cooling zone is not very clear.

In the present study, a three-dimensional model coupling fluid flow and heat transfer in the mold has been developed to compute the strand temperature field under different mold corner structure: right-angle, big-chamfer, multi-chamfer and fillet. Then the temperature field at mold exit was exported to the two-dimensional heat transfer model in secondary cooling zone to compute the corner temperature field. Based on the simulation results, genetic influence of mold corner structure on the strand corner temperature in secondary cooling zone was discussed during slab continuous casting.

Mathematical Model

3-D Fluid Flow and Heat Transfer Model in the Mold

(a) Assumptions

In the mathematical model, the liquid steel in mold was assumed to be a steady-state, three-dimensional, incompressible Newtonian fluid. The strand curvature, bulging, oscillation, mold taper and air gap were neglected. Besides, the heat transfer at the strand free surface was ignored.

(b) Fluid and Turbulence Model

The liquid steel flow in the liquid pool was achieved through solving the continuity and momentum equations. The popular K-ε turbulent model [7] was chosen. The equations are:

$$\rho \frac{\partial(\rho u_j)}{\partial x_i} = 0 \tag{1}$$

$$\rho \frac{\partial u_i u_j}{\partial x_j} = \frac{\partial}{\partial x_j}\left[\mu_{eff}\left(\frac{\partial u_i}{\partial u_x} + \frac{\partial u_j}{\partial x_i}\right)\right] - \frac{\partial P}{\partial x_i} + S_P \tag{2}$$

where u_i and u_j is the velocity component, x_i and x_j is the spatial coordinate, ρ is the density of liquid steel, P is the pressure, μ_{eff} is the effective viscosity, S_P is the sink term of velocity.

(c) Solidification and Heat-Transfer Model

The governing equation for heat transfer and solidification is:

$$\rho u_i \frac{\partial H}{\partial x_i} = \frac{\partial}{\partial x_i}\left(\lambda_{eff} \frac{\partial T}{\partial x_i}\right) \tag{3}$$

where λ_{eff} is the effective thermal conductivity, T is the temperature, H is the enthalpy of steel, which is the sum of the sensible enthalpy and latent heat.

(d) Computational Domain

Due to the symmetry of fluid flow and heat transfer, only a quarter of the strand was modelled to minimize computation. The casting parameters are shown in the Table 1. Figure 1 is the mesh of the strand and the four kind of mold corner structure: right-angle, big-chamfer, multi-chamfer and fillet. The strand was meshed using about 820,000 hexahedral cells.

(e) Boundary Condition

At the inlet of SEN, the velocity-inlet boundary condition was used. The inlet velocity was calculated through the mass conservation based on the casting speed. The inlet value of turbulent kinetic energy and the rate of turbulent energy dissipation was 10^{-5} [9], respectively. The pressure-outlet boundary condition was applied at strand exit.

The top free surface was set as zero shear stationary wall and heat insulation. At the symmetry plane, the normal velocity components and normal gradients of all other variables were assumed to be zero. At the strand surface, the heat transfer boundary condition was employed the heat flux, which decreased along the casting direction.

Table 1 The casting parameters of mold [8]

Operating parameters	Values
Mold section	2000 mm × 260 mm
Mold length	800 mm
Computer length	2800 mm
Inside size of SEN (submerged entry nozzle)	80 mm × 60 mm
Outside size of SEN	140 mm × 140 mm
Port size of SEN	60 mm × 95 mm
Port angle	−15°
Casting speed	1.15 m/min
Casting temperature	1803 K

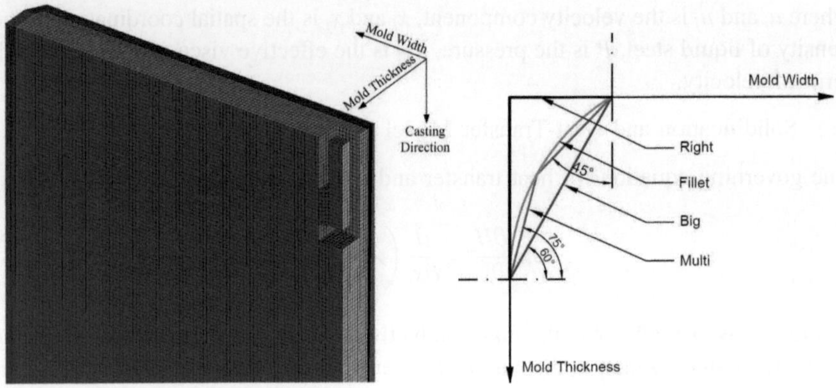

(a) The mesh of the strand (b) Mold corner structure

Fig. 1 The schematic diagram of mold and corner structure

2-D Heat Transfer Model in Secondary Cooling Zone

(a) Assumptions

The liquid steel flows along the casting direction was ignored in this model. Therefore, the simulation of strand in secondary cooling zone was simplified to a two-dimensional heat transfer model. The latent heat of solid phase transformation was ignored as well.

(b) Computational Domain

The 2-D model is presented in Fig. 2. It is also a quarter model in consideration of the symmetry. This model moves at casting speed from mold exit to secondary cooling zone exit to compute the heat transfer during the whole secondary cooling process.

(c) Boundary Condition

The heat flux boundary condition was applied at strand surface, which included the heat transfer of water impact, radiation and roller contact. According to actual continuous casting process, the strand was divided into 13 regions along the casting

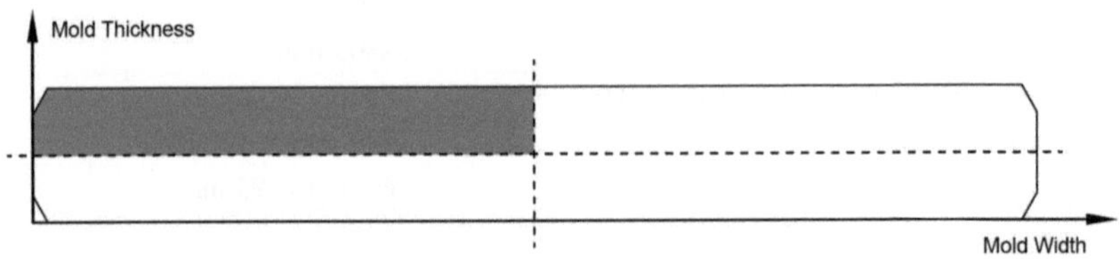

Fig. 2 The schematic diagram of the 2-D model

Table 2 The physical properties of steel

Physical properties	Values
Density (kg/m^3)	7200
Specific heat (J·(kg·K)$^{-1}$)	700
Thermal coefficient (W·(m·K)$^{-1}$)	61.46−0.02 × T(K)
Viscosity (kg·(m·s)$^{-1}$)	0.0062 [10]
Latent heat (J/kg)	264300
Solidus temperature (K)	1736
Liquidus temperature (K)	1787

direction. In addition, the strand was divided into 5 regions along the width direction to simulate the non-uniform secondary cooling water density.

Physical Properties

In the simulation, the physical properties of steel are presented in Table 2.

Solution Procedure

In this study, the CFD software FLUENT 15.0 was employed. The 3-D model was solved at a time step of 0.05 s for 350 s to make sure all residuals were below 1×10^{-3} and the average temperature on the outlet was steady. Then the temperature field at mold exit was exported to be the initial condition of the 2-D model. The 2-D model was solved at a time step of 0.5 s for 1740 s to compute the whole secondary cooling process.

Results and Discussion

Temperature Distribution on the Slab Surface

Figure 3 shows the temperature field of strand surface under different mold corner structure: (a) right-angle, (b) big-chamfer, (c) multi-chamfer and (d) fillet. It is obvious that the temperature distribution on the wide face of the four kind of strand is very close. The mold corner structure has little influence on the temperature field on wide face. The temperature distribution on narrow face is a little different from that on wide face. On the lower part of narrow face, the temperature distribution is non-uniform due to the nozzle jet. Under the influence of corner structure, the temperature near mold exit of big-chamfered strand is about 50 K higher than that

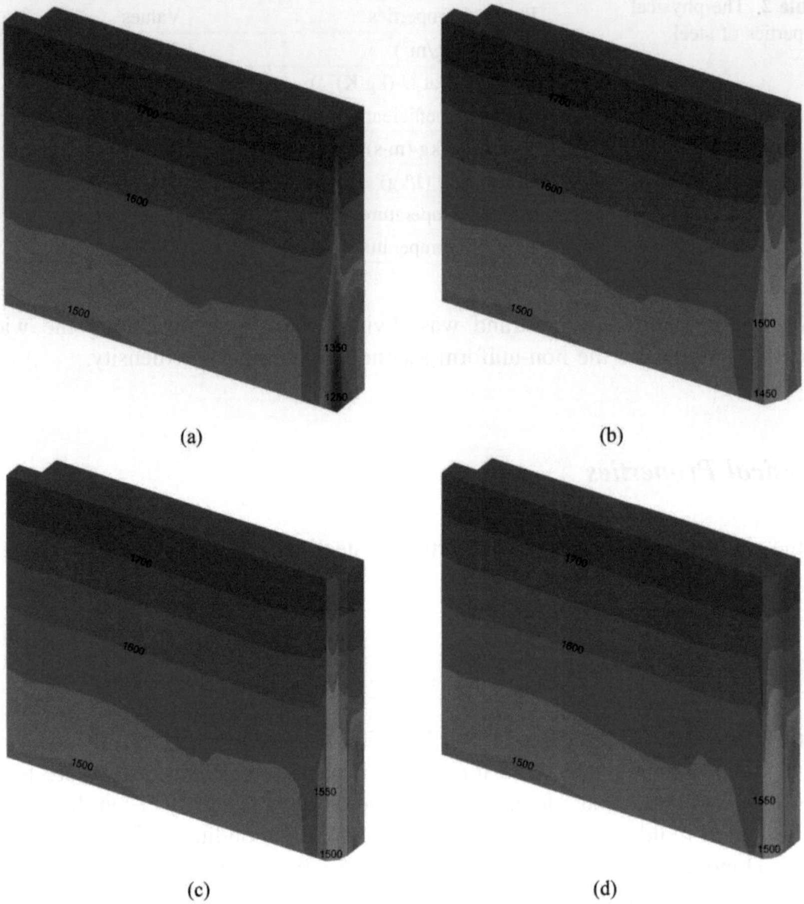

Fig. 3 The temperature field of strand surface under different mold corner structure: **a** right-angle, **b** big-chamfer, **c** multi-chamfer, **d** fillet

of right-angle strand. And the temperature of multi-chamfered and fillet strand is about 50 K higher than that of big-chamfered strand as well.

On the chamfered face, the temperature distribution presents significant differences. The temperature around the corner is lower than that of wide face and narrow face. The corner temperature of right-angle strand is the lowest because of the 270° heat transfer. The lowest corner temperature of right-angle, big-chamfered, multi-chamfered and fillet strand is 1224, 1437, 1447 and 1447 K, respectively. It means that the using of chamfered mold will promote the corner temperature obviously. While the corner structure changed from big-chamfer to multi-chamfer, the rise of corner temperature is relatively small. Moreover,the corner temperature of multi-chamfered strand and fillet strand is almost the same.

Effect of Mold Corner Structure on the Strand Corner Temperature

Figure 4 compares the cooling trend of strand corner temperature among the four kind of strand. In the figure, "w" means the corner between the wide face and the chamfered face, while "n" means the corner between the narrow face and the chamfered face. And "c" means the corner between the two chamfered faces in multi-chamfered strand. It is evident that the corner temperature of the right-angle strand is the lowest in secondary cooling zone. On one hand, the right-angle structure will make the cooling faster because the heat transfer angle of right-angle strand is larger than that of chamfered strand. On the other hand, the corner temperature of right-angle strand at mold exit is much lower than that of chamfered strand. Besides, when corner structure changed from right-angle to big-chamfer, the rise of corner temperature was much higher than that changed from big-chamfer to multi-chamfer or fillet. Therefore, it is very effective to improve the corner temperature in secondary cooling zone through using chamfered mold. Although multi-chamfered mold and fillet mold may increase the corner temperature further, it is not remarkable relatively.

The corner temperature variation in straightening segment is shown in Fig. 5. The entry and exit corner temperature is presented in Table 3. It is clear that the corner temperature of right-angle strand in straightening segment is between 1037 K and 1064 K, which is in the low ductility temperature zone. The strand may have corner cracks under the influence of thermal stress and straightening stress. The corner temperature of chamfered strand is about 100 K higher than that of right-angle strand, which will avoid the low ductility temperature zone. It is very useful to prevent corner cracks. Moreover, when corner structure changed from big-chamfer to multi-chamfer or fillet, the rise of temperature is less than 20 K.

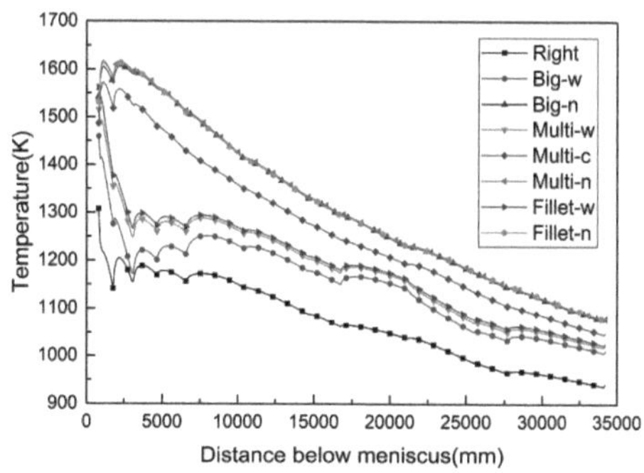

Fig. 4 Strand corner temperature along the casting direction in secondary cooling zone

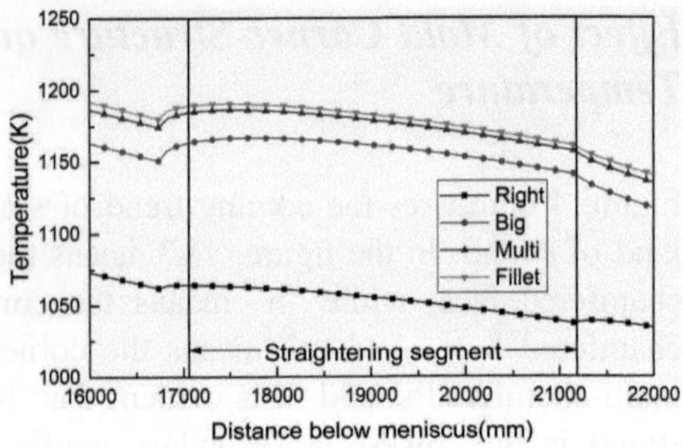

Fig. 5 Strand corner temperature along the casting direction in straightening segment

Table 3 The entry and exit corner temperature of straightening segment

Corner structure	Entry temperature (K)	Exit temperature (K)
Right-angle	1064	1037
Big-chamfer	1163	1141
Multi-chamfer	1184	1157
Fillet	1189	1161

Strand Surface Temperature Along the Strand Cross Section

Figure 6 is the strand surface temperature along the cross section at straightening segment entry. In the figure, "0" is the corner between the wide face and narrow face of right-angle strand, while the corner between the wide face and chamfered face of chamfered strand. The temperature difference around the corner of the right-angle strand is much larger than that of chamfered strand. It may result in large thermal stress. And the temperature difference of the big-chamfered strand is

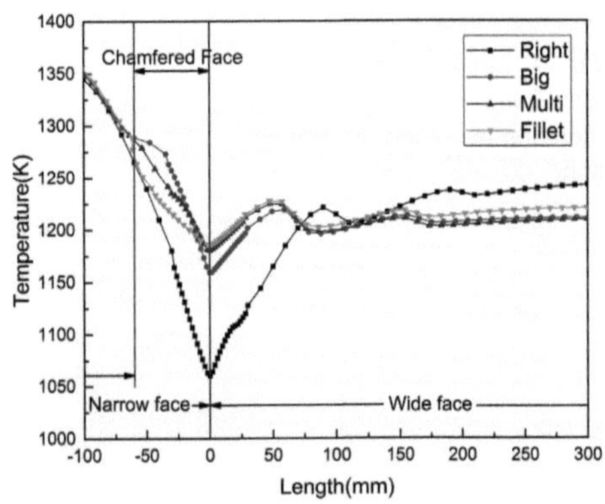

Fig. 6 Strand surface temperature along the cross section at straightening segment entry

larger than that of multi-chamfered and fillet strand. The temperature of multi-chamfered and fillet strand has a relatively smooth shift in the strand surface. It is helpful to decrease the appearance of corner crack. The difference between the multi-chamfered strand and fillet strand is that the temperature of fillet strand in the chamfered face is lower than that of multi-chamfered strand.

In addition, both from Figs. 4 and 6, it can be seen that for the three kind of chamfered strand, the corner temperature between the wide face and chamfered face is obviously lower than that between the narrow face and chamfered face. On one hand, the angle of the corner between the wide face and chamfered face is smaller than that between the narrow face and chamfered face. On the other hand, the temperature of narrow face is higher than that of wide face near mold exit owing to the impact of nozzle jet. Besides, the secondary cooling water density on narrow face is much smaller than that on wide face.

Strand Central Temperature Under Different Mold Corner Structure

Figure 7 gives the strand central temperature change along the casting direction during the secondary cooling process. In the figure, "Right" means right-angle strand, "Big" means big-chamfered strand, "Multi" means multi-chamfered strand, "Fillet" means fillet strand. It shows that the central temperature of the right-angle strand is higher than that of chamfered strand and it solidifies fastest. There is little difference among the three kind of chamfered strand. The final solidification point of right-angle strand, big-chamfered strand, multi-chamfered strand and fillet strand is 22.1, 20.5, 20.4 and 20.2 m, respectively. It is clear that chamfer will speed up the heat transfer rate of the strand because of the increase of heat transfer area.

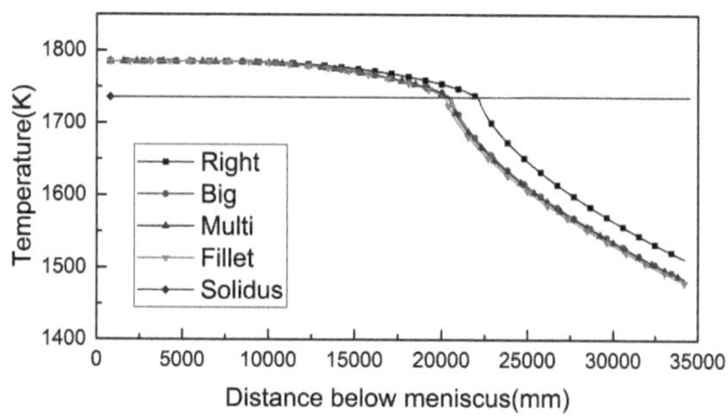

Fig. 7 Strand central temperature along the casting direction in secondary cooling zone

Conclusions

In this work, a three-dimensional model coupling fluid flow and heat transfer in the mold and a two-dimensional heat transfer model in secondary cooling zone was established to research slab corner temperature under different mold corner structure. Conclusions are as follow:

(1) The corner temperature at mold exit will be increased about 200 K through using chamfered mold. When the corner structure changed from big-chamfer to multi-chamfer or fillet, the rise of corner temperature is relatively small.
(2) In secondary cooling zone, the corner temperature of chamfered strand is about 100 K higher than that of right-angle strand, which will avoid the low ductility temperature zone during the straightening process. The corner temperature of big-chamfered strand, multi-chamfered strand and fillet strand increases in turn, but the increase amplitude is relatively small.
(3) The strand surface temperature of multi-chamfered and fillet strand along the strand cross section shifts relatively smoothly, which will decrease the thermal stress.

Acknowledgements The work is financially supported by the Natural Science Foundation of China, Project No. 51374260 and 51611130062.

References

1. Mintz B, "Importance of Ar3 temperature in controlling ductility and width of hot ductility trough in steels, and its relationship to transverse cracking," *Mater Sci Tech-Lond*, 12(2) (1996), 132–8.
2. Brimacombe J, Sorimachi K. "Crack formation in the continuous casting of steel," *Metallurgical transactions B*, 8(2) (1977), 489–505.
3. Patrick B, Ludlow V. "Development of casting practices to minimise transverse cracking in microalloyed steels," *La Revue de Metallurgie-CIT Juilet-Aout*, (1994), 1081–1089.
4. WANG Wen-jun et al., "Development and Application of Mould with Chamfered Corners for Continuous Casting of Slabs," *Journal of Iron and Steel Research*, 24(9) (2012), 21–26.
5. CAO Jian-xin et al., "Application of Chamfered Mould on Slab Continuous Casting Production in Lianyuan Iron and Steel Company," *Iron and Steel*, 48(11) (2013), 43–47.
6. Ren Fei-fei et al., "Numerical Simulation of Actual Temperature Field for Chamfered Mould Copper," *Iron and Steel*, 50(4) (2015), 27–33.
7. Yang H et al., "Mathematical simulation on coupled flow, heat, and solute transport in slab continuous casting process," *Metallurgical and Materials Transactions B*, 29(6) (1998), 1345–1356.
8. Xie X et al., "Mathematical Modeling of Heat Transfer in Mold Copper Coupled with Cooling Water During the Slab Continuous Casting Process," *Metall Mater Trans B*, 45(6) (2014):2442–2452.

9. Chaudhary R et al., "Transient Mold Fluid Flow with Well- and Mountain-Bottom Nozzles in Continuous Casting of Steel," *Metall Mater Trans B*, 39(6) (2008), 870–884.
10. Najera-Bastida A et al., "Shell Thinning Phenomena Affected by Heat Transfer, Nozzle Design and Flux Chemistry in Billets Moulds," *Isij Int*, 50(6) (2010), 830–838.

Author Index

A. Allanore et al. (eds.), *Materials Processing Fundamentals 2017*,
The Minerals, Metals & Materials Series, DOI 10.1007/978-3-319-51580-9

Subject Index

A. Allanore et al. (eds.), *Materials Processing Fundamentals 2017*,
The Minerals, Metals & Materials Series, DOI 10.1007/978-3-319-51580-9

MIX
Papier aus verantwortungsvollen Quellen
Paper from responsible sources
FSC® C105338

If you have any concerns about our products,
you can contact us on
ProductSafety@springernature.com

In case Publisher is established outside the EU,
the EU authorized representative is:
Springer Nature Customer Service Center GmbH
Europaplatz 3, 69115 Heidelberg, Germany

Printed by Libri Plureos GmbH
in Hamburg, Germany